Gravitational Field - Concepts and Applications

Edited by Khalid S. Essa

Published in London, United Kingdom

IntechOpen

Supporting open minds since 2005

Gravitational Field - Concepts and Applications
http://dx.doi.org/10.5772/intechopen.94782
Edited by Khalid S. Essa

Contributors
Ahmad Alhassan, Auwal Aliyu, D.M Dahab Abdelfattah, Francis T.S. Yu, Khalid S. Essa, Zein E. Diab, Mukaila Abdullahi, Ahmed Shehab Ahmed Al-Banna, Vladimir Stepanovich Korolev, Elena Polyakhova, James F. Woodward

Notice
Statements and opinions expressed in the chapters are these of the individual contributors and not necessarily those of the editors or publisher. No responsibility is accepted for the accuracy of information contained in the published chapters. The publisher assumes no responsibility for any damage or injury to persons or property arising out of the use of any materials, instructions, methods or ideas contained in the book.

First published in London, United Kingdom, 2022 by IntechOpen
IntechOpen is the global imprint of INTECHOPEN LIMITED, registered in England and Wales, registration number: 11086078, 5 Princes Gate Court, London, SW7 2QJ, United Kingdom
Printed in Croatia

British Library Cataloguing-in-Publication Data
A catalogue record for this book is available from the British Library

Additional hard and PDF copies can be obtained from orders@intechopen.com

Gravitational Field - Concepts and Applications
Edited by Khalid S. Essa
p. cm.
Print ISBN 978-1-83969-752-4
Online ISBN 978-1-83969-753-1
eBook (PDF) ISBN 978-1-83969-754-8

Meet the editor

Dr. Khalid S. Essa received his BSc with honors (1997), an MSc (2001), and a Ph.D. (2004) in geophysics from the Faculty of Science, Cairo University, Egypt. He joined the staff of Cairo University in 1997 and was appointed a research professor of potential field methods in the Department of Geophysics in 2014. He was a visiting post-doctoral researcher at the University of Lorraine and Strasbourg University, France in 2018; Charles University, Prague in 2014; and Western Michigan University, USA in 2006. Dr. Essa is an editor and external reviewer for many top journals. He has attended several international geophysical conferences in the United States, Australia, and France. He is a member of the Society of Exploration Geophysicists (SEG), American Geophysical Union (AGU), American Association of Petroleum Geologists (AAPG), European Association of Geoscientists and Engineers (EAGE), and Egyptian Syndicate of Scientific Professions (ESSP). He is also a member of the Petroleum and Mineral Resources Research Council and the National Committee for Geodesy and Geophysics. In 2017, he received the Prof Nasry Matari Shokry Award in Applied Geology from the Academy of Scientific Research and Technology and an Award for Scientific Excellence in Interdisciplinary, Multidisciplinary and Future Sciences from Cairo University.

Contents

Preface

The study of gravitational fields is important for learning about the Earth and space. Sir Isaac Newton's law of universal gravitation, which states that the force between two masses is proportional to their product and inversely proportional to the square of their separation distance, opened the door to the development and investigation of gravity mathematical theories. The study of gravity allows for measuring the mass distribution and, as a result, deducing the internal structure and shape of the planets.

This book discusses gravitational fields in eight chapters organized into two sections: "Some Concepts Concerning the Gravity Method" and "Application of the Gravity Method in Exploration."

Section 1 consists of four chapters that demonstrate the theory of gravity and discuss gravitational fields. Chapter 1 by Al-Banna discusses Newton's law and its principal concepts; gravitational potential and gravitational attraction; geoid, spheroid, and geodetic figures of the Earth; and the gravity difference between the equator and poles. Chapter 2 by Woodward presents the results of an experiment designed to generate transient gravitational forces at a practical level in the laboratory. Chapter 3 by Yu shows the nature of our temporal (t > 0) universe because without temporal space there would be no gravitational force and a gravitational field cannot be created within an empty space. The chapter also examines how gravitational waves can be created. Chapter 4 by Polyakhova and Korolev highlights the main tasks of photo-gravitational celestial mechanics and the possibilities of their mathematical modeling.

Section 2 also consists of four chapters. Chapter 5 by Abdullahi reviews the benefits of gravity surveys in exploration and discusses the use of spectral analysis to design a filter for the separation of residual and regional anomalies of the complete Bouguer anomaly and its interpretation. Chapter 6 by Essa and Diab utilizes an R-parameter imaging technique to interpret a gravity anomaly profile. This technique depends on estimating the correlation factor between the analytic signal of the real gravity anomaly and the analytic signal of the forward gravity anomaly of the assumed buried source. Chapter 7 by Abdelfattah presents a new 2D semi-inversion method for delineating and tracing the thickness of formations and the depth of basement rocks for a deposition basin using gravity data. Finally, Chapter 8 by Alhassan and Aliyu demonstrates how to apply the second vertical derivative method for satellite gravity data of a location in Nigeria to enhance weaker local anomalies, define the edges of geologically anomalous density distributions, and identify geologic units.

This book is a useful resource for scientists, researchers, physicists, and geophysicists wishing to delve deeper into the study of gravity and gravitational fields.

Khalid S. Essa, Ph.D.
Professor,
Faculty of Science,
Geophysics Department,
Cairo University,
Cairo, Egypt

Some Concepts Concerning the Gravity Method

Chapter 1

Gravity Field Theory

Ahmed Shehab Al-Banna

Abstract

Gravity keep all things on the earth surface on the ground. Gravity method is one of the oldest geophysical methods. It is used to solve many geological problems. This method can be integrated with the other geophysical methods to prepare more accepted geophysical model. Understanding the theory and the principles concepts considered as an important step to improve the method. Chapter one attempt to discuss Newton's law, potential and attraction gravitational field, Geoid, Spheroid and geodetically figure of the earth, the gravity difference between equator and poles of the earth and some facts about gravity field.

Keywords: gravitational theory, gravitational attraction, ellipsoid and geoid, gravity variation with latitude, facts about gravity field

1. Introduction

The theory of gravitational method based on Newton's law expressing the force of mutual attraction between two particles in term of their masses and separation.

This law states: (that two very small particles of mass (m_1) and (m_2) respectively, each with dimensions very small compared with the separation (r) of their centers of mass, will be attracted to one another with a force.

$$F = G\frac{m_1 m_2}{r^2} \tag{1}$$

G = Universal gravitational constant
667×10^{-8} cm^{3}g^{-1}s^{-2} in cgs system
6.67×10^{-11} (Nm^2/Kg^2) in SI system
If m_1, m_2 in gram and r (cm)
The second law of motion expressed mathematically by the following. F = ma
Where
m – the mass a = acceleration,
r measured by (cm), M_1, m_2 measured in (gm), F measured by (dyne).

2. Gravitational acceleration

The acceleration (a) of a mass (m_2) due to the attraction of a mass (m_1) a distance (r) away can be obtained simply by dividing the attracting force F by the mass (m_2).

$$F = \frac{Gm_1 m_2}{R^2} \tag{2}$$

$$g = a = \frac{F}{m_2} = \frac{Gm_1}{R^2} \tag{3}$$

The acceleration is the conventional quantity used to measure the gravitational field acting any at point [1].

Is the Earth's gravitational acceleration is constant? No, it is not, that is due to the variation of mass distribution and variation of the diameters of the Earth.

In the cgs system, the dimension of acceleration is (cm/sec^2). Among geophysicists this unit is referred as the (Gal) (in honor of Galileo, who Conducted pioneering research on the earth's gravity).

The gravitational acceleration at the earth surface is about (980 cm/sec^2) or (980 Gal), but in exploration work we are likely to be measuring differences in acceleration. 10^{-7} of the earth's field. The unit which is more convenience in working with gravity data for geological and geodetic studies is milli-gal (milligal) or

$$(mGal) = \frac{1}{1000} \text{ gal} = 10^{-3} \text{ gal}$$

This unit has come to be the common unit for expressing gravitational accelerations.

There is another unit called gravity Unit (g.u.) which is equal to 0.1 mGal.

g.u. = 0.1 milligal = 10^{-4} gal.

and micro- gal = 10^{-6} gal.

3. Gravitational potential

The intensity of gravitational field depends only on position, the analysis of such fields can often be simplified by using the concept of potential.

The potential at a point in a gravitational field is defined as the work required for arbitrary reference point to the point in question.

The acceleration at a distance (r) from (p) is

$$= Gm/r^2 \tag{4}$$

The work necessary to move the unit mass a distance (ds) having a component (dr) in the direction of (p) is

$$= (Gm_1/r^2) \, dr \tag{5}$$

The work (U) done in moving the mass from infinity to a point (O), (**Figure 1**) in the gravitational field of (m_1) is

$$U = Gm_1 \int^R dr/r^2 \tag{6}$$

$$U = Gm_1 \frac{1}{r} \Big|_a^R = \frac{-Gm_1}{R} \tag{7}$$

$$U = Gm/R \tag{8}$$

The quantity (Gm1/R) is the gravitational potential. It is depend only on the distance (R) from the point source (m_1). By differentiating both sides of the above equation it can be seen that the gravitational acceleration is the derivative of the potential with respect to (r).

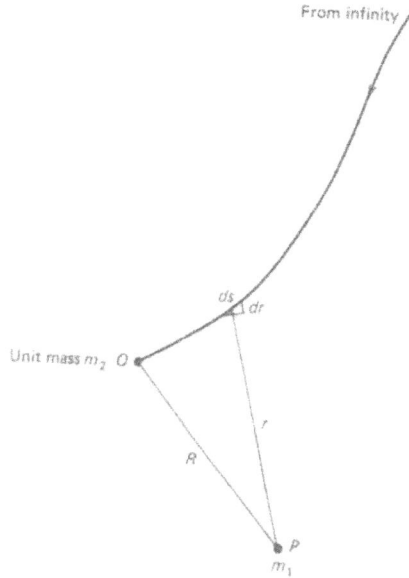

Figure 1.
The movement of mass unit from infinity to the point (O) [1].

Any surface along with the potential is constant, so it is referred as an equipotential surface.

Sea level, for example, is an equipotential surface, even though the actual force of gravity varies along the sea surface by more than (0.5%) between the equator and either of the poles.

4. Newton's law and large dimensions mass

When the dimensions of the source are large, it is necessary to extend the theory. The procedure is to divided the mass into many small elements, and to add the effects of each of these elements, together to measure the gravity effect on certain point.

Because force or acceleration is a vector having both magnitude and direction, it is necessary to resolve the force from each element of mess into its three components (most generally its vertical component and its north–south and east–west components in horizontal plane), before the attraction of the body at any point can be determined.

If we consider the attraction of an irregular laminar body (part of two-dimensional sheet) in the xz plane at an external point (p). We first determine the x (horizontal) and Z (vertical) components of acceleration at (p) associated with this attraction.

To do this we divide the plate into (N) small elements of mass, each of area (Δs), if the density (ρ_n) is uniform within the (n^{th}) elements we express the (x) component of celebration at point (p) due to the attraction of this element as, (**Figure 2**). laminar body, by divided the mass into small masses (Δs) [1].

$$g_{xn} = \frac{G\rho_n \Delta S}{r_n^2} \cos\theta = \frac{G\rho_n \Delta S}{r_n^2} \cdot \frac{X}{r_n} = \frac{GX}{r_n^3} \rho_n \Delta S \qquad (9)$$

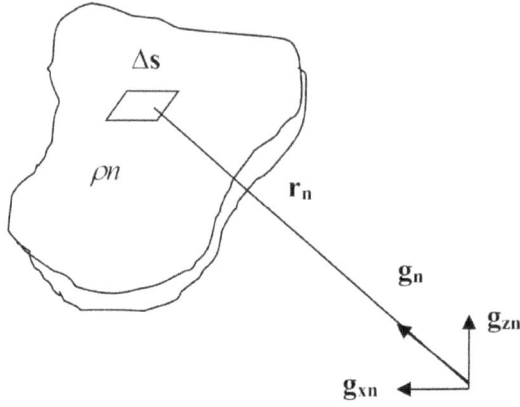

Figure 2.
The gravity determination of two dimensional irregular.

and

$$g_{zn} = \frac{G\rho_n \Delta S}{r_n^2} \sin\theta = \frac{G\rho_n \Delta S}{r_n^2} \cdot \frac{Z}{r_n} = \frac{GZ}{r_n^3} \rho_n \Delta S \tag{10}$$

Adding the acceleration for all elements

$$g_x = G \sum_1^N \frac{X\rho_n}{r_n^3} \cdot \Delta S \tag{11}$$

$$g_z = G \sum_1^N \frac{Z\rho_n}{r_n^3} \cdot \Delta S \tag{12}$$

Where ΔS is very small we can express the two components of acceleration at (p) by the respective integration.

$$g_x = G \int^S \frac{\rho_x X}{r^3} ds \tag{13}$$

$$gz = G \int^S \frac{\rho_z Z}{r^3} ds \tag{14}$$

Where S = area of body
P = (mass per unit area).
This case can be extended to three dimensional case for a mass per unit volume.

$$g_x = G \int^V \frac{\rho_x X}{r^3} dv \tag{15}$$

$$g_y = G \int^V \frac{\rho_y y}{r^3} dv \tag{16}$$

$$g_z = G \int^V \frac{\rho Z}{r^3} dv. \tag{17}$$

Where (V) is the volume of the body [1].

In gravity exploration, only the vertical component of force is measured, so that we are normally concerned only with (g_z) in determining the attraction at the surface of a buried body.

One of the most important properties of Potential, that it is satisfy Laplace equation anywhere outside the effective gravity mass.

$$\nabla^2 A = \frac{\partial^2 A}{\partial x^2} + \frac{\partial^2 A}{\partial y^2} + \frac{\partial^2 A}{\partial z^2} = O \tag{18}$$

Laplace equation caused the Ambiguity in gravity and magnetic fields.

5. Gravitational attraction

Newton's law states that in case of two masses (m_1) and (m_1) separated by a distance (r) between them, then these two masses attract each other's by force (F).

$$F = \frac{Gm_1 m_2}{r^2} \tag{19}$$

G = Universal gravitational constant

If The mass of the earth (M), and the radius of the earth (r), then the weight (w) of a body with a mass (m) on the surface of the earth equal to:

$$W = mg = \frac{GMm}{r^2} \tag{20}$$

The above equation used only for non-rotating earth.

$$g = \frac{GM}{r^2} \tag{21}$$

(g) usually expressed as a weight of a unit mass or earth attraction.

The gravity attraction of the earth varied from point to point on the earth surface, due to that the radius of the earth (r) (which is not constant everywhere on the earth) in addition to the centrifugal force create duo to the rotation of the earth, (**Figure 3**).

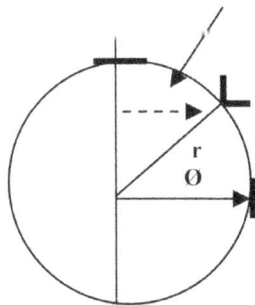

Figure 3.
The variation of centrifugal force value with latitude.

The gravity attraction (g) can be measured with high accuracy and the size of the earth can be determined using the astronomical geodetic studies. And as the universal gravitational constant (G) is known, the earth mass can calculated easily.

The gravity attraction can be measured using a pendulum method with accuracy of 5 ppm, while the relative variation in gravity can be measured using special high accurate gravity meter instruments (composed of springs system) with an accuracy of 10^{-9}.

The following equation used on assumption of non-rotated earth

$$F = \frac{GMm}{r^2} \tag{22}$$

But the earth is a rotated body. It is rotate around the long axis. This rotation created a centrifugal force (a) varied relatively with the variation of the radius of the circular shape plotted by a rotated points on the earth surface.

$$a = r\omega^2 \tag{23}$$

Where
ω is the angular velocity
T is the period

$$\omega = \frac{2\lambda}{T} \tag{24}$$

Then

$$\alpha = 4\pi^2 \, r/T^2 \tag{25}$$

So
α value at the equator is 3.4 gal

$$a = 3.4cm/\sec^2 = \frac{g}{289} \tag{26}$$

For the other latitude circle of the earth which symbolized by Ø, the radius (r) replaced by

$$r \cos Ø$$

then the angular acceleration become

$$\alpha \cos Ø$$

This acceleration have two components:
The vertical component is

$$\alpha \cos^2 Ø$$

This component reach its maximum value at the equator (3.4 gal), while its value at the pole is zero which is the minimum value.
The horizontal component is

$$\alpha \cos Ø \sin Ø,$$

The horizontal component reach its maximum value (0.5 gal) at the latitude Ø = 45°, while it reach its minimum value (zero) at the equator and poles.

The total gravity force which is determine the weight of anybody on the earth during a free fall represented as a resultant of the gravitational attraction of the earth and the centrifugal force at certain point on the earth.

Tenth thousands of gravity measurements on the earth surface indicate that g values at poles is greater than g values at equator by about (1/189) of the total gravity value. Where the centrifugal force caused a difference between the equator and poles by about (1/289), while the increase in radius at the equator by (21 km.) more than pole caused a variation by about (1/547).

1/289 = 0.003460207 Centrifugal force effect.

1/547 = 0.00182815 Earth radius difference effect

!/189 = 1/289 + 1/547

1/189 = 0.0052884 The difference in gravity between the equator and poles relative to the total gravity value.

6. Gravitational theory

The first approximation of the shape of the earth is the sphere.

The second approximation of the earth is the oblate spheroid.

For theoretically studies and for simple applications it is possible to use the horizontal surface (level surface) or (equipotential surface everywhere, which is perpendicular on the plumb line (force line).

The problem of determine the shape of the earth is actually is the problem of determine the shape of equipotential surface.

The gravitational field include infinity equipotential surfaces.

These equipotential surfaces are not intersect at all.

The scientists deals to considered the sea level as the reference equipotential surface in gravitational studies and call this surface geoid after Listing 1873.

The equipotential surface not necessarily coincide with the equal gravity surface. So the sea level considered approximately a surface of equipotential gravitational surface (because it is perpendicular on the gravity force at every point).

7. Clairaut theorem

At 1743 the French mathematical scientist Clairaut found a mathematical expression to represent the relation between the gravity measurements on the earth surface and the shape of the earth.

Supposing that.

a = maximum radius of oblate spheroid.

b = minimum radius of oblate spheroid.

a and b are the major semi axes of an oblate spheroid of revolution. and write

$$b = a\,(1 - f). \qquad (27)$$

$$f = (a–b)/a \qquad (28)$$

F = oblateness or ellipticity of the spheroid for spherical body

$$((x^2 + y^2)/a^2) + (z^2/(a^2\,(1–2\,f))) = 1 \qquad (29)$$

That is

$$x^2 + y^2 + z^2\,(1 + 2f) = a^2 \qquad (30)$$

if
is the colatitude angle Θ
the above equation written as

$$r^2 \left(1 + 2f \cos^2 \Theta\right) = a^2 \qquad (31)$$

or

$$a^2/r^2 = \left(1 + 2f \cos^2 \Theta\right) \qquad (32)$$

Considering the earth as an oblate spheroid, rotate around its axis and the variation in gravity measurements on the earth, Clairaut obtained the following theorem:-

$$g = g_e \left[1 + \left(\frac{5}{2}C - f\right) \sin^2 \Phi\right] \qquad (33)$$

Where
g = Theoretical gravity value at any point on the earth according to its latitude angle.
g_e = gravity at the equator.
Φ = latitude angle

$$C = \omega^2 a/g_e = (\text{Centrifugal force/Attraction force}) \text{ at equator} \qquad (34)$$

(where latitude angle equal zero)
The variable of $\sin^2 \Phi$ which is
$(5c/2) - f$
Represent a gravitational flattening (β)
Which is written in other expression as following:

$$\beta = \frac{g_p - g_e}{g_e} \qquad (35)$$

g_e = Gravity at equator
g_p = Gravity at pole
The Clairaut's theorem also written in other expression:

$$g_. = g_e \left(1 + \beta \sin^2 \Phi\right) \qquad (36)$$

The value of the factor β can be obtained from measuring a lot of absolute gravity values at different locations at the earth surface. The slope of the best fit line of the gravity value versus the latitude angle represent the factor β multiplied by the gravity at the equator g_e.

$$\text{Considering} f = 1/297 \qquad (37)$$

The determination of gravity values at the sea level over the world led to obtained the following equation:

$$G = 978.049\left(1 + 0.0052884 \sin^2 \Phi - 0.0000059 \sin^2 2\Phi\right) \qquad (38)$$

Grand and West, [2].

10

This equation called the international gravity formula, or the 1939's equation, where the gravity value at equator (g_e) is (978.049 gal), which calculated statistically from measurements on the earth surface. The value of the factor of ($sin^2\Phi$) represent the flattening factor effect, while the value of the factor of ($sin^2 2\Phi$) represent a correction value to fit the earth shape with rotated spheroid body shape. These two factors depends on the shape of the earth and speed of rotation of the earth.

Depending on the principles of gravity anomalies the earth seems as triaxle ellipsoid. The long axis of the earth lying 10° west of Greenwich and the difference between radiuses of the earth may be within 150 ± 58 meters, which is considered very low variation relative to the average radius of the earth therefore it is neglected in most cases.

There is another formulas such as

Helmert, 1901 formula

$$g_o = 978.030 \left(1 + 0.005302 \; sin^2 \Phi - 0.000007 \; sin^2 2\Phi\right) \tag{39}$$

the radius of the according to this equation are:

a = 6378200 m.

b = 6356818 m

and

$$f = 1/298.2 \tag{40}$$

This formula used in old gravity measurements.

Other formula used in 1917 in the United States of America using the following formula:

$$g_o = 978.039 \left(1 + 0.005294 \; sin^2 \Phi - 0.000007 \; sin^2 2\Phi\right) \tag{41}$$

The above formula is obtained depending on 216 gravity stations in USA, 42 in Canada, 17 in Europe, and 73 gravity stations in India. The flattening value of this formula (1/297.4).

The international gravity formula 1967 which is adopted by the Geodetic Reference System (GRS-1967) with different factors values according to new observations [3]

$$g_o = 978.03185 \left(1 + 0.0053024 \; sin^2 \Phi - 0.0000059 \; sin^2 2\Phi\right) \tag{42}$$

This formula used to remove the variation in gravity with latitude (latitude correction). This formulae consider the earth as rotating ellipsoid without geologic or topographic complexities.

The GRS 1980 formula is

$$g_o = 978.0327 \left(1 + 0.0053024 \; sin^2 \Phi - 0.0000058 \; sin^2 \Phi\right) \tag{43}$$

The gravity survey for small sites for examples in case of engineering studies the latitude correction approximated using the a correction factor 0.813 sin 2Φ mgal/ km in the north- south direction.

8. Geocentric latitude

The geocentric latitude represent the angle between a line at any point at the earth surface and passing through the center of the earth and the plane of the equator.

9. Geodetic latitude (Φ)

The geodetic latitude is the geographical latitude. The geodetic is the angle between normal line on the geoid (which is approximately the earth shape) at any point on the earth and the plane of equator (which approximately the earth shape). The geodetic latitude is differ from geocentric latitude because of elliptical shape of the earth.

The geodetic latitude - geocentric latitude = 11.7 sin 2Φ in minutes of arc. The maximum value difference between them is about 21.5 km. at latitude 45°.

10. Geoid

It is an gravitational equipotential surface. The vertical gravity component is perpendicular on this surface on all its points. This surface is approximately equal to the sea surface in seas and oceans. In land area it is considered as the extension of sea level below the continents.

11. The normal spheroid, ellipsoid and geoid

The earth considered as an ideal spheroid or normal spheroid in case of considering the earth as a completely liquid, means without any lateral density change. The direction of gravity attraction in such case is perpendicular on the earth everywhere and passing through the earth center. But the increase of radius of the earth at equator from that at poles by about 21 km make the ellipsoid shape is better approximation for earth than spheroid shape, (**Figure 4**).

Also, the earth in fact is not uniform and the density change laterally at least in the crust and upper most part of the mantle of the earth. The actual surface of the earth can be represent by the geoid which is equal to sea level and its extension in the land. The geoid surface may be up or down the ellipsoid surface depending on the distribution of density in the earth or the topographic changes, (**Figure 5**). The evidences on the difference between the geoid and ellipsoid obtained from the observation of the deflection of plumb line. This deflection measured during the geodetic and astronomical measurements, where the plump line deflect toward the excess masses in the continent, while it is deflected away from an area of mass deficiency as in oceans area.

The difference between geoid and ellipsoid surfaces small relative to the radius of the earth. For example it is about 40 meters in Rocky Mountains.

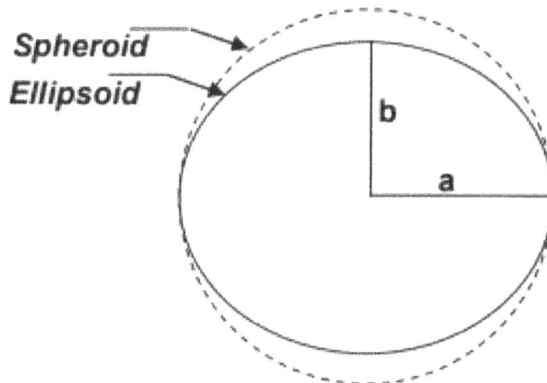

Figure 4.
The spheroid and ellipsoid relationship [4].

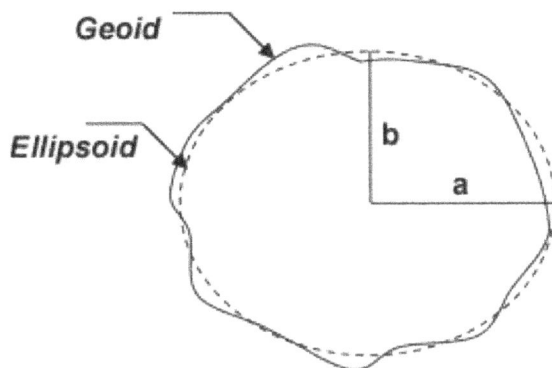

Figure 5.
The ellipsoid and geoid relationship [4].

12. Line of forces

The gravity field can be described by lines. These lines usually perpendicular on the equipotential surface. When the body isotropic, these lines is vertically on the surface. These lines coincide with the force direction of the gravity field. If these lines are converging to each other that means an access of mass or positive anomaly, the diverging of these lines indicate a deficiency in mass or negative anomaly.

13. Component of force

The three basic principle directions of force are:

$$Fx \rightarrow x$$
$$Fy \rightarrow y$$
$$Fz \rightarrow z$$

The equation of force lines are

$$\frac{\partial x}{Fx} - \frac{\partial y}{Fy} = \frac{\partial Z}{Fz} \tag{44}$$

The force lines which pass a unit area (ds) vertically on the surface of the mass expressed as a Gravity Field Intensity

$$g = \frac{\Phi}{ds} \tag{45}$$

where Φ is labeled "Lines No." and ds is labeled "Unit area".

All points in the space outside the attractive mass in the potential gravity field characterized its subject to Laplace equation. The fact is the reason of ambiguity in gravity field.

$$\nabla^2 A = \frac{\partial^2 A}{\partial x^2} + \frac{\partial^2 A}{\partial y^2} + \frac{\partial^2 A}{\partial z^2} = 0 \tag{46}$$

Where A is the potential field which is function of point position according to (x), (y), (z)

14. Gradient of potential

The gradient of potential define as the force divided by the mass

$$\text{gradient of potential} = \frac{-Force}{mass} \tag{47}$$

If i, j, k are vector units, and the gravity field is U

$$g = \left[\vec{i}\frac{du}{dx} + \vec{j}\frac{\partial u}{\partial x} + \vec{k}\frac{\partial u}{\partial z} \right] \tag{48}$$

∇ = del $\Rightarrow$ Gradient operator

$$g = \vec{\nabla}U$$
$$\vec{\nabla} = \left[\vec{i}\frac{\partial}{\partial x} + \vec{j}\frac{\partial}{\partial y} + \vec{k}\frac{\partial}{\partial z} \right] \tag{49}$$

Gauss's law:- The total gravitation flux through any closed surface is equal to $(-4\lambda G)$ times the mass enclosed by the surface.

$$\Phi = -4 \ \lambda GM \tag{50}$$

15. Gauss's law for force flux

The surface integral of the normal component of intensely of the gravitational field gives the flux through the closed surface.

$$\vec{\Phi} = \int \vec{g}.\vec{ds} \tag{51}$$

$\vec{ds}$ = vector normal to the surface (ds) and magnitude equaled to the area (ds). The integral over the whole surface gives Gauss's law for total mass enclosed by the surface.

(The surface integral $\Rightarrow$ volume integral) divergences theorem

$$\vec{\Phi} = \int_5 \vec{g}.\vec{ds} \Rightarrow \int_v \vec{\nabla}.\vec{g}.dv \tag{52}$$

$$\vec{\nabla}.\vec{g} = \frac{\partial g x}{\partial x} + \frac{\partial g_y}{\partial y} + \frac{\partial g_z}{\partial z} \tag{53}$$

$$\therefore \phi = \int_v \vec{\nabla}.\vec{g} \ .dv = -\int_v \vec{\nabla}\nabla U dv = -4\lambda GM \tag{54}$$

$$= \int_v \overrightarrow{\nabla^2} U.dv = +4\lambda GM \tag{55}$$

$$= \int_v \overrightarrow{\nabla^2} U.dv = +4\lambda G \int_v \sigma dv \tag{56}$$

Where σ = The density of mass distribution in the volume V
Poisson's equation

$$\overrightarrow{\nabla}.\overrightarrow{\nabla} U = 4\lambda G \sigma \tag{57}$$

Laplace equation of the potential (U) of the gravitational field is equal to a constant times the density of the distribution matter in the field.

16. Laplace theorem

For a point Located outside attracting masses. The sum of second order derivation of the attraction potential along the axes of orthogonal coordinate is

$$\frac{\partial^2 u}{\partial x^2} + \frac{\partial^2 u_y}{\partial y^2} + \frac{\partial^2 u}{\partial z^2} = 0 \tag{58}$$

But when the point being attracted lies inside the attracting mass the Laplace operator becomes

$$\Phi = -4\pi G \ \sigma \tag{59}$$

Simply we can be considered that Laplace equation is a special application of Poisson's equation, where the mass density is zero, This case found outside the attracting mass (outside the earth surface).

17. Forces of gravity

In addition to the attraction force of the earth mass, there is another force which effect the rotated earth called (Centrifugal Force). This force created due to the continuous rotation of the earth around its axis. The force related to the radius of rotation and the square of the angular velocity:-
Therefore
The total gravity force = Attraction Force + Centrifugal Force (FC)

$$Fc_p = r\omega^2 COS\theta \tag{60}$$

P = point at the earth surface.
Θ = latitude
Centrifugal force
$Fc \ p \ \alpha \ r \ \omega^2$

$$Fcp = mr\omega^2 \tag{61}$$

Where m = mass

r = radius of rotation
ω = angular velocity

$$\therefore -g = -G \int \frac{1}{r^2} dm + mr\omega^2 \tag{62}$$

The Centrifugal force contribute the total gravity effect on the earth surface by about 10^{-7}–10^{-8} gal. The force have its maximum value at the equator of the earth and reach zero value at the poles of the earth.

The absolute gravity value at the equator is about 978 gal, while its value at poles about 983 gal, The difference in gravity value between the equator and pole is range between (5.17 gal) or (5170 mgal).

18. The difference in gravity between equator and pole

1. The acceleration of centrifugal force act outward (away from the earth) in other word opposite of gravity attraction. The centrifugal force reach maximum value at equator and its minimum value, which equal zero at poles. The factor create a difference in gravity between the equator and poles of about 3.39 gal. So this factor reduce the gravity at equator by 3.39 gal. Therefore the gravity seems more at poles by **+3.39 gal** than at equator.

2. The Poles are nearer to the center of the earth than the equator by about 21 km. This factor will increase the gravity at poles than at equator by about **+6.63 gal**.

3. The mass-shape of the earth (increase of earth radius at equator) will cause an increase of gravity attraction at the equator than that at the pole due to an increase of mass, by about 4.85 gal. Therefore the gravity at poles seems lower than at equator by - **4.85 gal**.

Finally the summation of gravity value difference indicated that the pole gravity is more at poles than the equator by about +5.17 gal. According to the following equation
The gravity at poles = + 3.39 + 6.63–4.85 = + 5.17 gal (The excess in gravity value at the pole relative to the equator) (63).

19. Some facts about gravity field

1. Gravity method involves measurement a field of force in the earth that is neither generated by the observer nor influence by anything he does.

2. The field of gravitational or magnetic prospecting is a composite of contributions can be individually resolved only in special care.

3. In gravity measurements, the quantity actually observed is not the earths true gravitational attraction but its variations from point to another.

4. The instruments are designed to measure difference in gravity rather than its actual magnitude.

5. The variation of gravity depend only upon lateral changes in density of earth materials in the vicinity of the measuring point, [5].

6. The density is a physical property that changes significantly from one rock type to another. Knowledge of the distribution of this property within the ground would give information, of great potential value about the subsurface geology.

7. Density is the source of a potential field which is intrinsic to the body and acts at a distance from it. The strength of gravitational field of a body is in proportion to its density.

8. All materials in the earth influence gravity, but because of the inverse square law of behavior of rocks that lie close to the point of observation will have a much greater effect than those farther away.

9. The bulk of the gravitational pull of the earth (i.e. the weight of a unit mass), however, has little to do with the rocks of the earth's crust. It is caused by the enormous of the mantle and core, and since these are regular in shape and smooth varying in density so the earth's gravitational field is, in the main regular and smoothly varying also.

10. Only about three parts in one thousand (0.3%) of (g) are due to the material contained within the earth's crust, and of this small amount roughly (15%) 0.05% of (g) is accounted for by the uppermost (5 km) of rock (that region of the crust generally being the base of geological phenomena). Changes in the densities of rocks within this region will produce variations in (g) which generally do not exceed (0.01%) of absolute gravity value anywhere.

11. The geological structures contribute very little to the earth gravity, but the importance of that small contribution lies in the fact that it has a point to-point variation which can be mapped.

12. To produce meaningful maps, two imperatives must be fulfilled these are:

    A. The measuring apparatus must be sufficiently sensitive to detect the effect of geology on (g).

    B. Effective methods must be used to compensate the data for all sources of variation other than the local geology. These corrections will includes chiefly the effect of changing elevation and of crustal heterogeneity on abroad scale

13. When data are finally reduced to a form meaningful in terms of local geology, they must be interpreted.

The interpretation of gravity field is done mainly in two parts:
First solution is sought to the (inverse) potential problem, which Consists in deducing the shape of the source from its potential field these solutions are never unique.
Second, The solutions deduced from (potential field theory) are interpreted in geological terms. This requires some knowledge of the factors which determines the densities of rocks.

20. Some selected useful terms of gravity field modified

1. **Density contrast:-** The density difference between the gravity source and the host rocks [6]. In case of excess of mass it is positive, While it is negative in case of mass deficiency.

2. **Ellipticity:-** The ratio of the major to minor axes of an ellipse.

3. **Eötvös unit:-** A unit of gravitational gradient which is equal to 10^{-6} mgal/cm.

4. **Graticule:** - A template for graphically calculating of gravity.

5. **Gravimeter:** - An instrument for measuring variation in gravitational attraction (gravity meter).

6. **Hilbert – Transform technique:-** A technique for determining the phase of a minimum- phase function from its power spectrum.

7. **Inverse problem:** The problem of gaining knowledge of the physical features of a disturbing body by making observations of its effects, (finding the model from the observed data).

8. **Direct (Forward) (Normal) Problem:** Calculating the possible suspected values from a given model.

9. **Invers square law:** A potential field surrounding a unit element has a magnitude inversely as the square of the distance from the element (in gravity case the element is the massand the field is equal to Gm/r^2).

10. **Sensitivity:** The least change in a quantity which a detector is able to perceive, (An instrument can have excellent sensitivity and yet poor accuracy).

11. **Accuracy:** The degree of freedom from error (the total error compared to the true value).

12. **Readability:** The least discernible change in a readout device, which can be readily estimated.

13. **Precision:** The repeatability of an instrument (measured by the mean deviation of set of readings from the average value).

14. **Repeatability:** The maximum deviation from the average of corresponding data taken from repeated tests under ideal conditions.

15. **Tidal effect:** Variation of gravity observation due to the distribution of the earth resulting from attraction of the moon and sun.

Gravity Field Theory
DOI: http://dx.doi.org/10.5772/intechopen.99959

Author details

Ahmed Shehab Al-Banna
Department of Geology, College of Science, University of Baghdad, Baghdad, Iraq

*Address all correspondence to: geobanna1955@yahoo.com

IntechOpen

References

[1] Dobrin, M. B. 1976. Introduction to geophysical prospecting. McGraw-Hill, New York. 3rd edition. 630p.

[2] Grant F. S. and West, G. F. 1965. Interpretation theory in applied geophysics. McGraw–Hill, New York. 584p.

[3] Kearey, p. , Brooks, M. and Hill, I. 2002. An Introduction to geophysical Exploration. Blackwell Publishing, 3rd edition. 262p.

[4] Al-Sadi, H. N. and Baban E. N. 2014. Introduction to gravity exploration method. Paiwand press, University of Sulaimani University, Sulaimaniyah, Kurdistan Region, Iraq. 1st edition. 205p.

[5] Nettleton, L. L. 1971. Elementary gravity and magnetic for geologist and seismologists. Society of Exploration Geophysics. 121p.

[6] Sheriff, R. E. 1976. Encyclopedic Dictionary of Exploration Geophysics, Society of Exploration Geophysicists, Tulsa, Oklahoma. 3rd edition, 266p.

Chapter 2

Gravity and Inertia in General Relativity

James F. Woodward

Abstract

The relationship of gravity and inertia has been an issue in physics since Einstein, acting on an observation of Ernst Mach that rotations take place with respect to the "fixed stars", advanced the Equivalence Principle (EP). The EP is the assertion that the forces that arise in proper accelerations are indistinguishable from gravitational forces unless one checks ones circumstances in relation to distant matter in the universe (the fixed stars). By 1912, Einstein had settled on the idea that inertial phenomena, in particular, inertial forces should be a consequence of inductive gravitational effects. About 1960, five years after Einstein's death, Carl Brans pointed out that Einstein had been mistaken in his "spectator matter" argument. He inferred that the EP prohibits the gravitational induction of inertia. I argue that while Brans' argument is correct, the inference that inertia is not an inductive gravitational effect is not correct. If inertial forces are gravitationally induced, it should be possible to generate transient gravitational forces of practical levels in the laboratory. I present results of a experiment designed to produce such forces for propulsive purposes.

Keywords: gravity, inertia, general relativity, inertia as a gravitationally induced phenomenon, experimental test of inductive inertia

1. Introduction

Before Einstein's creation of general relativity theory, the conception of inertia was that captured in Newton's laws of mechanics, Newton's elaboration of the idea of inertia, first introduced by Galileo some years earlier. Inertia, properly *vis inertiae* or inert force, was taken to be an inherent property of "matter" conferred on it by its existence in absolute space, that only ceases to be inert when external forces act on matter to produce proper accelerations, rising to produce the reaction force the matter exerts on the accelerating agent to resist the impressed force. Already in Newton's day, this conception of inertia as due to absolute space was seriously called into question by, among others, Bishop Berkeley who argued that a body in an otherwise empty universe would have no inertia since there would be no other matter to refer motion of the body to. From this point of view, absolute space's action on matter is not the origin of inertia, the action of other matter in space is the cause of inertia. In the 17th and 18th, and most of the 19th centuries, Newton's view prevailed.

Berkeley's conjecture was revisited in the late 19th century by Ernst Mach, who noted that local rotation coincided with rotation relative to the "fixed stars", suggesting that local inertial frames of reference were determined by some long-range action of matter at cosmological distances. Einstein took Mach's insight to mean that in any properly constituted theory of gravity, inertia would emerge as an "inductive" gravitational effect of cosmic matter since gravity was/is the only known long-range force that might cause such effects. His first explicit attempt in this direction appeared in his, "Is There a Gravitational Effect Which is Analogous to Electrodynamic Induction?" in 1912 [1]. Einstein noted that Newtonian gravity was not sufficient to correctly encompass the induction of inertia – that is, the generation of the mass of matter by the gravitational interaction with chiefly cosmological matter – and inertial reaction forces – that is, Newton's third law forces on accelerating agents. Induction requires vector or tensor interactions. Several years later, general relativity was Einstein's theory that he was convinced accomplished this task. Indeed, that's why he called it "general relativity" because he, as we would say today, "unified" inertia and gravity by making inertia an inductive gravitational effect. Analogous to Maxwell's "unification" of electricity and magnetism in his electrodynamics.

2. Einstein's conception of gravity and inertia in general relativity

Einstein started talking about the gravitational induction of inertia as "Mach's principle" shortly after mooting general relativity. Willem de Sitter quickly pointed out that the field equations of general relativity have solutions that are plainly inconsistent with any reasonable interpretation of "Mach's principle". Einstein retreated from full-blown Mach's principle, which seemed to require action at a distance, then deemed inconsistent with the conception of field theory as articulated by Faraday. But he did not abandon the gravitational induction of inertia which is consistent with the tenets of field theory. Einstein advanced his ideas first in an address at Leiden in 1920 where he analogized his evolving view of spacetime to the "aether" of the turn of the century theory of electrodynamics. That is, spacetime is not some pre-existing void in which matter, gravity and the other forces of nature exist. It is a real, substantial entity – not a void – which *is* the gravitational field of matter sources. And then he extended his view in remarks in a series of lectures at Princeton in 1921 [2]. He calculated the action of some nearby, "spectator" matter on a test particle of unit mass (at the origin of coordinates) in the weak field limit of GR. There he found for the equations of motion of the test particle (his Equations 118):

$$\left(\frac{d}{dl}\right)[(l + \bar{\sigma})\mathbf{v}] = \nabla\sigma + \frac{\partial \mathbf{A}}{\partial l} + \nabla \times (\mathbf{A} \times \mathbf{v}), \tag{1}$$

$$\bar{\sigma} = (\kappa/8\pi) \int (\sigma/r)\ dV_o, \tag{2}$$

$$\mathbf{A} = (\kappa/2\pi) \int (\sigma d\mathbf{x}/dl) r^{-1} dV_o. \tag{3}$$

The second and third of these equations are the expressions for the scalar (σ) and vector (A) potentials of the gravitational action of the spectator masses with density σ on the test particle. l is coordinate time and v is coordinate velocity of the test particle. The first equation is just Newton's second law. After writing down these equations, Einstein noted approvingly that,

> *The equations of motion, (118), show now, in fact, that.*
> *The inert mass [of the test particle of unit mass] is proportional to 1 + σ, and*
> *therefore increases when ponderable masses approach the test body.*

> There is an inductive action of accelerated masses, of the same sign, upon the test
> body. This is the term dA/dl …
> Although these effects are inaccessible to experiment, because κ [Newton's constant
> of universal gravitation] is so small, nevertheless they certainly exist according to the
> general theory of relativity. We must see in them strong support for Mach's ideas as
> to the relativity of all inertial interactions. If we think these ideas consistently
> through to the end we must expect the whole $g_{\mu\nu}$-field, to be determined by the
> matter of the universe, and not mainly by the boundary conditions at infinity.

Note that the small effects singled out by Einstein – the contribution of gravitational potential energy to the rest mass of the test particle and the inductive action, a force, of accelerating masses – are the two features of gravity that would supposedly account for all of inertia – origin of mass and reaction forces – in a properly constituted cosmology.

The above quote was not Einstein's last explicit word on gravity, inertia, and spacetime. In 1924, he again addressed these topics in a paper, "Concerning the Aether" [3]. In it he quickly asserted that by "aether" he did not mean the material aether of turn of the century electromagnetism. Rather, he meant a real, substantial, but not material entity that *is* spacetime, and that spacetime *is* the gravitational field of material sources. No material sources, no spacetime. Why did he make this radical break with the conception if space as a pre-existing void in which nature plays out its events in time? Arguably, this was his way of getting rid of the Minkowski and other metrics that de Sitter had shown to be anti-Machian, delimiting acceptable solutions of his field equations to those consistent with inertia as a strictly gravitational interaction. As he put it toward the end of his article:

> The general theory of relativity rectified a mischief of classical dynamics. According
> to the latter, inertia and gravity appear as quite different, mutually independent
> phenomena, even though they both depend on the same quantity, mass. The theory of
> relativity resolved this problem by establishing the behavior of the electrically neu-
> tral point-mass by the law of the geodetic line, according to which inertial and
> gravitational effects are no longer considered as separate. In doing so, it attached
> characteristics to the aether [spacetime] which vary from point to point, determin-
> ing the metric and the dynamical behaviour [sic.] of material points, and deter-
> mined, in their turn, by physical factors, namely the distribution of mass/energy.
>
> That the aether of general relativity differs from those of classical mechanics and
> special relativity in that it is not "absolute" but determined, in its locally variable
> characteristics, by ponderable matter. This determination is a complete one if the
> universe is finite and closed …

One may reasonably ask, if Einstein was convinced that general relativity, correctly interpreted, encompassed the gravitational induction of inertia, why today is it widely believed in the community of relativists and beyond that inertia is not gravitationally induced? That inertia is no better understood now than it was in the absolute systems of Newton and Minkowski? Carl Brans. And his "spectator matter" argument.

3. Carl Brans' "spectator matter" argument

Brans did his doctoral work at Princeton in the late 1950s. His doctoral supervisor was the noted experimentalist, Robert Dicke. After passing his qualifying exam,

Dicke tasked Brans with investigating the question of as the origin of inertia in general relativity, as Dennis Sciama and others had made "Mach's principle" a central question in general relativity several years earlier. When Brans read Einstein's remarks on Machian inertia in Einstein's 1921 comments quoted above, he noted a problem. If gravitational potential energy due to nearby matter contributes to the rest masses of test particles, the Equivalence Principle is violated. This is not a minor problem. The Equivalence Principle is the bedrock of general relativity. The solution to this problem adopted in Brans' graduate school days was the imposition of a "coordinate condition". (A "gauge" solution is/was not available as general relativity is not a gauge theory).

As Brans, responding to several published papers on Mach's principle, later wrote in 1977 [4]:

> Over the years, many and varied expressions of Mach's principle have been pro-
> posed, making it one of the most elusive concepts in physics. However, it seems clear
> that Einstein intended to show that locally measured inertial-mass values are grav-
> itationally coupled to the mass distribution in the universe in his theory. For conve-
> nience I repeat the first order geodesic equations given by Einstein to support his
> argument:
> [Brans inserted here Einstein's equations displayed above.]
>
> ... Einstein's claim is that "The inertial mass is proportional to $(l + \bar{\sigma})$, and there-
> fore increases when ponderable masses approach the test body.

Brans pointed out that having the masses of local objects, the unit mass test particle in this case, depend on their gravitational potential energies acquired by interaction with spectator matter must be wrong. Were it true, then the electric charge to mass ratios of elementary particles for example would depend on the presence of nearby matter. If this were true, gravity could be discriminated from accelerations without having to check for the presence of spectator matter by going to the window in a small lab and looking out to see if one were on Earth, or in a rocket accelerating at one "gee" in deep outer space– a violation of the Equivalence Principle. From this, Brans inferred that

> ...global, i.e., nontidal, gravitational fields are completely invisible in such local
> standard measurements of inertial mass, contrary to Einstein's claim ... Einstein
> ought to have normalized his local space-time measurements to inertial frames, in
> which the metric has been transformed approximately to the standard Minkowski
> values, and for which distant-matter contributions are not present. [Emphasis
> added.]

This is the "coordinate condition" required by Brans' work: that the coordinates be compatible with the assumed approximate Minkowski metric appli-cable in small regions of spacetime. Since the absence of gravity is presupposed for Minkowski spacetime, this amounts to the assumption that the Newtonian potential due to exterior matter in such small regions of spacetime is effectively everywhere/when equal to zero. That is, the locally measured value of the total Newtonian gravitational potential is universally zero. This certainly makes the localization of gravitational potential energy impossible in general relativity, a now widely accepted fact. And where there is effectively no gravity, there can be no gravita-tional induction of inertia. Accordingly, it would seem that Brans' spectator matter argument makes Machian gravitationally induced inertia incompatible with general relativity.

Carl Brans and his then graduate advisor Robert Dicke created scalar-tensor theory to redress this perceived failing of general relativity. A clear indicator of the importance of the spectator matter argument. The eventual failure of scalar-tensor gravity has left the question of the nature of inertia in a limbo that remains unresolved. Nonetheless, the spectator matter argument still stands as the key issue in the development of general relativity in the past century as it requires that any theory of inertia must satisfy the Equivalence Principle. A test that has not yet been met for any theory of gravity, general relativity included if Brans' argument is accepted as completely correct.

4. Is inertia gravitationally induced in general relativity?

The problem of the origin of inertia, that is, Mach's principle, ceased to be a topic of mainstream interest in general relativity 50 years ago. It was not forgotten by those who lived through the '50s and '60s and were parties to the debates on inertia. For example, John Wheeler, working with Ignazio Ciufolini, made it the center piece of one of his last major books on gravity: *Gravitation and Inertia* [5]. On the otherwise blank page facing the first page of chapter 1 we find, "Inertia here arises from mass there". Exactly as Einstein would have said. In the penultimate chapter (5) of the book we find, "In the next chapter we shall describe in detail dragging of inertial frames and *gravitomagnetism*, which may be thought of as a manifestation of some weak general relativistic interpretation of the Mach principle, their measurement would provide experimental foundation for this general relativistic interpretation of the origin of inertia."

Brans' spectator matter argument does not involve gravitomagnetism. But it is implicitly present in Einstein's Eqs. 118 quoted above. In the vector potential **A**. The sources of **A** are the matter density currents in the universe. As Sciama pointed out in his first paper "On the Origin of Inertia" [6], the integration over cosmic matter currents involved can be vastly simplified by noting that the important currents involved can be singled out by assuming the local accelerating body in question and (instantaneously) has velocity **v**, can be taken as at rest with the universe moving past it rigidly with velocity − **v**. This can be removed from the integration, and the remaining integral just returns the Newtonian gravitational potential for all the stuff in the universe (up to a factor of order unity). If the coefficient of the time derivative of − **v**, **a** that is, is one, then this term in the equation of motion is the inertial reaction force. So, Brans' argument does more than require that the gravitational potential energies conferred on test particles by spectator matter not influence the rest mass of the test particle. *It also demands that if inertial forces are gravitational inductive effects, the coefficient of the acceleration in the equation of motion must be one in all circumstances. This is only possible if the total Newtonian potential is a locally measured invariant equal to the square of the vacuum speed of light (which is also a locally measured invariant in general relativity).*

A series of events, too lengthy to relate in detail here, led to the rejection of the time derivative of the vector potential in the equation of motion in general relativity. This started with an article by Edward Harris in the *American Journal of Physics* in 1991 [7] where he outlined the analog of linearized weak field slow motion equations of general relativity with Maxwell's equations of electrodynamics. In this approximation, it is possible to argue that the $d\mathbf{A}/dt$ term in the field equation and equation of motion vanish by gauge invariance. Harris, of course, knowing that general relativity is not a gauge theory, allowed as how this could not be generally true. But others took this argument and tried to justify it on other terms. (See references in chapter 4 of Pfister and King. [8]) With the passing of the $d\mathbf{A}/dt$ term, so too went the question/possibility of the gravitomagnetic origin of inertial forces. This development led Herbert Pfister and Markus King, in their recent book *Inertia and Gravitation*, to remark that, "We hope to give a new synopsis of this theme [that Faraday induction that produces the time derivative of the vector potential is absent in general relativity], with this specific focus not entering most textbook presentations on the foundations of gravitation and general relativity – and in a way amend Cuifolini and Wheeler's view on gravitation and inertia … ".

Cuifolini and Wheeler had opted for an account of inertia based on an initial spacelike hypersurface complemented by elliptic (instantaneous) constraint equations (first discussed by Wheeler and independently Lynden-Bell in the '60s) in preference to integrations over matter currents out the past light-cone to the past particle horizon. Pfister and King, having rejected the existence of Faraday induction effects in general relativity, were left no choice but the hypersurface/constraint equation approach. All of this was, wittingly or unwittingly, motivated by Carl Brans' conclusion that in sufficiently small regions of spacetime one must use the Minkowski metric which makes the local Newtonian potential vanish, eliminating gravity from external sources as an actor at that scale. That is, making the presence of cosmic matter, in Brans' word, "invisible".

Who's right? Cuifolini and Wheeler? Pfister and King? Brans? Arguably, they are, in a sense, all right, and wrong. The source of the confusion and problems regarding inertia in general relativity is the Minkowski metric – which de Sitter showed Einstein to be an acceptable formal solution of his field equations. The Minkowski metric is the metric for flat pseudo-Euclidean spacetime with gravity completely absent. That it is a solution of Einstein's field equations is not surprising. The theory is constructed on the assumption that in sufficiently small regions, spacetime is flat (and special relativity applies). As Einstein explicitly claimed, however, without gravity, there is no spacetime. This makes the Minkowski metric an unphysical pre-general relativistic idealization. Scaffolding to be removed once the construction of the theory is complete. The Maxwellian analog is the "roller bearing" mechanical aether he used to construct his equations of electrodynamics, promptly abandoned once the construction was complete. But spacetime, as a matter of observation, is essentially flat almost everywhere/when. What takes the place of Minkowski spacetime in the completed theory? Spatially flat FLRW spacetime.

5. Spatially flat cosmology and spectator matter

The one thing in the discussion of the role of inertia in general relativity that is certainly right is that Brans was absolutely correct in asserting that if spectator matter contributes to the rest masses of test particles through its contribution to the total gravitational potential by changing it (as Einstein assumed), then the Equivalence Principle is violated. This *fact*, however, does not necessitate the assumption of Minkowskian spacetime in small regions. It simply requires that the total locally measured Newtonian gravitational potential is an invariant. If the Newtonian potential is everywhere exactly the same in local measurements, that, by itself, precludes charge to mass ratios of elementary particles depending on the local gravitational environment. The question then is, does the spatially flat spacetime of FLRW cosmology have the requisite properties for a spacetime/gravitational field that yields gravitational induction of inertia?

Spatially flat, that is, curvature index "k" = 0, cosmology has several remarkable properties beyond being the transitional case between spherical, closed and hyperbolic, open geometry. The metric for k = 0 FLRW cosmology is, aside from the scale factor which multiplies the spacelike part of the metric to produce cosmic expansion, time-independent. So, a universe that starts out spatially flat, stays spatially flat throughout its history. Brans' doctoral supervisor Robert Dicke, in the '70s, identified this as a paradox, for spatial flatness should be unstable against small perturbations that should drive cosmic spacetime curvature quickly into the spherical or hyperbolic state. Why, more than 10 billion years into cosmic expansion is the universe still spatially flat? Alan Guth created inflation to address this problem.

Curvature at cosmic scale is related to energetic considerations. This is especially clear in the Newtonian analog elaboration of cosmology. (See Bernstein's *An Introduction to Cosmology*, Prentice Hall, 1995, chapter 2 for example [9]). Spatial flatness is a consequence of the balancing of gravitational potential energy and "kinetic", that is, non-gravitational energy encompassed in the Eq. $E = mc^2$. If gravitational energy exceeds non-gravitational energy, the universe is "closed" and will eventually contract after reaching some finite size. When non-gravitational energy exceeds gravitational energy, the universe is "open" and expands forever. Since gravitational and non-gravitational energies are balanced in a spatially flat universe, if we place a test particle of mass m anywhere/when in spacetime (which is the gravitational field according to Einstein), we will have:

$$m_g\phi = m_i c^2 \qquad (4)$$

where the subscripts g and i identify the passive gravitational and inertial masses of the material particle. The Equivalence Principle identifies the passive gravitational and inertial masses as equal in magnitude, so:

$$\phi = c^2 \qquad (5)$$

everywhere/when. The vacuum speed of light, a constant in special relativity, becomes a locally measured invariant in general relativity because, while local measurements always return the same number for c, non-local measurements may return different numbers. (Distant observers measure c in the vicinity of black holes to be much less than their locally measured value). It follows that ϕ too must be a locally measured invariant like c in spatially flat cosmology. Is spacetime spatially flat at cosmic scale? Observation answers this question in the affirmative.

ϕ/c^2 being a ratio of locally measured invariants in k = 0 cosmology does two things. First, it means that Eq. (4) can be interpreted as the assertion that inertial mass is induced by the action of gravity due to cosmic sources, as Einstein claimed to be the case – notwithstanding Brans' spectator matter argument. Indeed, since Brans' argument can be sidestepped by ϕ being a locally measured invariant (equal to c^2), his argument becomes a compelling argument **for** the gravitational induction of inertial mass. Second, $\phi/c^2 = 1$ makes inertial reaction forces an inductive gravitational effect, again, as Einstein claimed should be the case. In this case, though, the claim is complicated by the tensorial nature of gravity.

In the vector approximation based on the analogy with electrodynamics, one writes for the "gravelectric" field equation:

$$E_{grav} = -\nabla\varnothing - \frac{1}{c}\frac{\partial A}{\partial t} \qquad (6)$$

in Gaussian units. If one uses this equation for gravity, when one computes the equation of motion for a test particle, one gets ϕ/c^2 times $dv/dt = \mathbf{a}$, the acceleration, from the term in the time derivative of the vector potential, and $\phi/c^2 = 1$ makes this term the inertial reaction force on the test particle. In tensor general relativity one must specify one's choice of coordinates.

6. Coordinate choice in general relativity

The most popular coordinate choice in general relativity is de Donder (harmonic) coordinates. When this choice is made, a factor of 4 appears in the term involving the analog, $g_{oi.}$, of the vector potential, A. This messes up the simple $\phi/c^2 = 1$ relationship between ϕ and c^2. But that relationship can be recovered by asserting that the gravelectric field equation be that just stated above [Eq. (6)] and only coordinates that return it are permissible. As did, for example, Braginski, Caves and Thorne [BCT] in their 1977 paper on "Laboratory experiments to test relativistic gravity" [10]. BCT, working with the coordinate choice of Misner, Thorne and Wheeler in chapter 39 of their massive *Gravitatjon* [11], worked through the details of the choice of Eq. (6) here for the potentials and metric. Their coordinate choice and gravelectric field equation determination leads to the correct Faradayan induction term in the equation of motion to account for inertial reaction forces. Edward Harris took partic- ular note of BCT's treatment of inductive effects before adopting the gauge invariance rejection of Faradayan induction in the weak field, slow motion, but time-dependent Maxwellian analog interpretation of general relativity that led to the creation of the now fashionable sub-discipline of "gravitoelectromagnetism"; so-called GEM theory. That led in turn to the rejection of Faradayan induction effects in general relativity generally noted by Pfister and King.

The demand of explicit gravitational induction of inertial reaction forces does more than simply limit one's choice of coordinates. It can also be used in conjunc- tion with Brans' spectator matter argument as a selection criterion for acceptable cosmologies. $k = \pm 1$ FLRW cosmologies, for example, do not conform to this criterion for in them, the gravitational and "kinetic" energies of test particles are not equal as $\phi \neq c^2$. If this is correct, then the remarkable *stability* of the $k = 0$ FLRW cosmology, remarked upon by Dicke and explained by Guth, is not a consequence of inflation. It is a consequence of the gravitational induction of inertial forces and Brans' spectator matter argument that makes ϕ a locally measured invariant equal to c^2.

What is important is that whatever one's choice in the matter is, that choice depends crucially on Carl Brans' spectator matter argument that, in turn, depends on the correctness of the Equivalence Principle – arguably the simplest expression of the *principle* of relativity for proper accelerations. If one chooses to go with Einstein regarding the role of inertia in general relativity, then Brans' argument dictates a coordinate choice like that of BCT. The question is: was Einstein right about the role of inertia in general relativity? That is a question that ultimately can only be answered by experiment.

7. Experiment and inertia

If one chooses to explain inertia as an inductive gravitational effect consistent with the EP, all one need do is impose a suitable coordinate condition. Then every real manifestation of inertia becomes, in a sense, an experimental demonstration of gravitational induction of inertia. But, in principle, all this has been known at least since BCT showed how to get the correct gravelectric field equation to account inductively for inertial forces. What we want is a novel experimental prediction that

Gravity and Inertia in General Relativity
DOI: http://dx.doi.org/10.5772/intechopen.99760

depends on the gravitational induction of inertia, a prediction that is not expected in the absence of inductive inertia. As it turns out, such a prediction exists, though it was not envisaged as a test of gravitational inertia induction when it was constructed. It is the prediction that if the local proper mass/energy density in a given region of spacetime is made to fluctuate, and that region is simultaneously subjected to a large proper acceleration to make manifest the inertial reaction gravitational field due to cosmic matter currents, that field vastly amplifies the magnitude of the masa/energy fluctuation.

This predicted rest mass fluctuation is a normally unobserved transient effect that is only obvious in special circumstances when it is sought. Those circumstances are the production of thrust in small systems, seemingly without the use of propellant, by coupling to the gravitational field of the universe through the mass fluctuations driven a stack of piezoelectric disks clamped to a brass reaction mass with an aluminum cap and screws. This is described in the precursor to this paper [12]. A device of this sort is show here in **Figure 1**. In order to maximize the oscillations of this device it is mounted using small linear ball bushings in ears on a flange on the reaction mass. The device is supported by rods in an aluminum frame as shown in **Figure 2**. A device like that in **Figure 1** has a mass of about 140 gms, whereas the support frame has a mass of about 60 gms. Work with devices like those shown in **Figures 1** and 2 commenced in the summer of 2020 using a high sensitivity torsion balance used in previous work. Large effects were produced. So large that the torsion balance was abandoned. But not until all of the various test for a genuine effect were completed.

The experiment was moved out of the balance vacuum chamber and onto a cantilever. Force generated in the device was measured by recording changes in the

Figure 1.
A Mach effect gravity assist (MEGA) impulse engine element. Eight 19 mm diameter by 2 mm thick led zirconium titanate disks are clamped between the aluminum cap and brass reaction mass. Linear ball bushings are fitted in the "ears" on the reaction mass.

Figure 2.
*A device like that in **Figure 1** mounted on steel rods in an aluminum frame. The device is centered in the frame on the rods by very soft springs that convey very low frequency and stationary forces to the frame without communicating any high frequency vibration.*

position of the device on the rods with a Philtec position sensor. Typical results with this arrangement are reported in [12].

The chief criticism that has been advanced of the Mach effects project is that the measured thrusts were not due to any real effect. Rather, they allegedly arose from simple vibration in the systems – so-called "Newtonian vibrational artifacts". Those of us working on the project, of course, had been careful to exclude such false positives. But those determined to believe that Mach effects do not exist persisted. Indeed, they still persist, notwithstanding that we have increased the forces generated by these devices by two to three orders of magnitude [13]. And this performance increase was achieved by isolating the strong vibrations in the device from the support structure using linear ball bushings in place of a simple rubber pad in earlier design devices. To quell lingering doubts about false positives arising from vibration, my partner in this work, Hal Fearn, resuscitated an antique air track to see if an air supported "glider" could be made to move thereupon. The results, for technical reasons, were equivocal. Hal then turned to a pendulum, made with a small plastic platform suspended by three fine nylon monofilament cords with length 1.9 m from the ceiling of our lab, This has been the force detection system in use for the past several months.

It quickly became apparent that the pendulum force detection system has one great advantage over all other force detection systems. It eliminates the significant inertia of the parts of other systems, For example, the significant mass of the beam and counter masses of a torsion balance disappear. All that remains is a few tens of grams for the plastic platform and adjustment screws added to the mass of the support structure. Why is this important? Well, the answer to the Newtonian vibrational artifact hypothesis, from the outset, has been that simple vibration induced in a system by the addition of energy, but no momentum, cannot produce a steady deflection of a force detection method by the conservation of momentum. Only the generation of a real force in the system can produce a steady deflection of force detection apparatus. The counter argument to this obviously correct momentum conservation argument is that induced vibration may produce a stick–slip mechanism in the parts of the system that result in the relative motion of parts of the system, and the motion of the part of the system attached to the force detection apparatus may displace the force sensor. While a transient displacement of a force sensor may result from such action, a steady displacement cannot occur for no steady, real force is generated by this process, and the restoring force of the force detection sensor will quickly re-zero the force sensor. Surprisingly, this obviously correct counter, counter argument has fallen on at least some deaf ears.

The beauty of the pendulum force detection scheme is that with large enough forces, the stick–slip scheme of the Newtonian hypothesis can easily be discriminated from the production of a real force. This is possible because the Mach effect forces in the new devices produce forces large enough (hundreds of micronewtons) to cause displacements of the pendulum on the order of hundreds of microns. The vibrations allegedly responsible for these displacements, however, have amplitudes less than a few hundred nanometers, known by direct observation with a Polytech laser vibrometer. The only way small amplitude vibrations can produce large amplitude displacements is by a stick–slip mechanism. And momentum conservation applied to this mechanism demands that the vibrating device and the support structure move in opposite directions with equal opposite momenta to preserve the location of the center of mass of the system as no net real force is generated. Since the mass of the support structure is roughly half the mass of the device, detection of these motions is a simple matter of simultaneous measurement of the positions of the device and support structure. If especially initially they move in opposite directions, you are looking at a Newtonian artifact. If they move together in the same direction, a real force is being generated in the device.

Figure 3 shows a device mounted on our pendulum platform with the vacuum chamber and torsion balance in the background in our lab. The positions of the device and support structure are measured with two Philtech position sensors as shown in **Figure 4**. The motion of the pendulum is only very lightly damped by the leads to the thermistor in the cap of the device that records the temperature of the device. In addition to the two position and temperature measurements, the voltage and current in the power circuit were monitored. Data were acquired and displayed with three Picoscopes and a Logitec BRIO webcam that captured the motion of the device in a movie displayed along with the Picoscope outputs, the entire screen being captured with software. A typical composite display screen is shown in **Figure 5**. The record of each run consists of the display screen capture movie and the strip-chart recording of the positions, voltage and temperature of one of the Picoscopes. The display screen movie is used to calibrate the position measurements in the strip-chart for conversion to force measurements.

The strip-chart recording for this run is shown in **Figure 6**. The fuzz on the gray trace of the support structure position has been post-acquisition cleaned up by a 10 Hz low pass digital filter and the temperature trace (green) has been added. Two

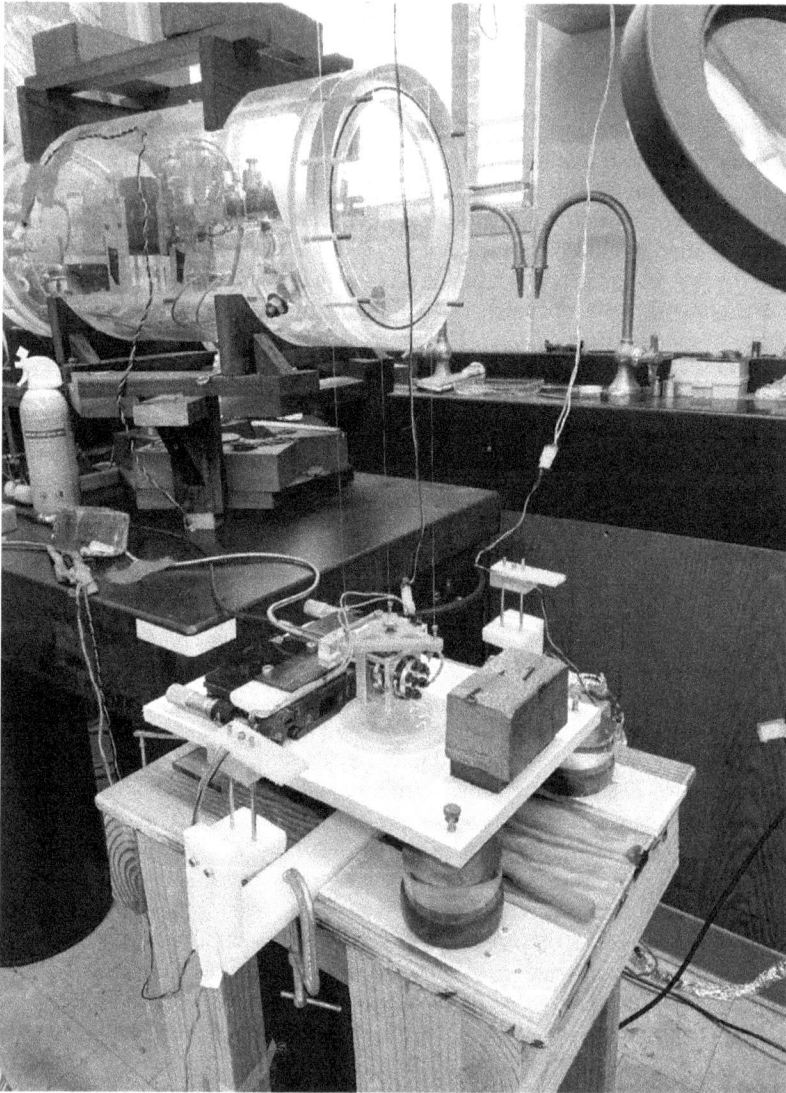

Figure 3.
The pendulum force sensing aparatus. The platform, the triangular green piece of plastic in the center of the picture, is suspended on three monofilament fibers attached at the ceiling of the room.

important inferences follow immediately from the data in **Figure 6**. First, since the two position sensors track together, it follows that a real force is generated in the MEGA impulse engine. The "level shifts" of the position traces are not consistent with "Newtonian vibrational ¯artifacts". Second, the prompt changes in the position traces at power on and off, allowing for some power on switching transient overshoot, indicate the presence of a steady force during the powered interval – as expected.

Resonances where Mach effect thrust is found, to date, have been located using the frequency sweep function of our Rigol signal generator, the present source of the single frequency sine function that drives a Carvin DCM-2000 power amplifier and 4 to 1 matching transformer. Once a suitable resonance is located, short 5 second constant frequency pulses are used to fine tune the optimal driving frequency.

Figure 4.
The position sensors are the two steel tubes on the left attached to micrometer stages.

This procedure will soon be supplanted by the automated routine on a Piezo Drives ultrasonic drivee that also supports resonance tracking.

The force that corresponds to the power-on displacement in **Figure 6** can be computed from the length of the pendulum, 1.85 m, the mass of the "bob", 0.23 kg, and the voltage to distance scale factor determined from the run movie, where for the device trace (red) a displacement of 0.5 mm correspond to a voltage change of 4.5 volt giving 0.11 mm per volt as the scale factor. The "level shift" from the incoming trace to the switching transient at power-on is about 2 volts, so the displacement produced by turning on the force is about 0.22 mm.. The vertical force on the pendulum is mg, or 2.3 newtons. The force that produces the 0.22 mm deflection is just the sine of the deflection angle, 2.2 X 10^{-4} m divided by 1.85 m, or 1.2 X 10^{-4}. This multiplied times the 2.3 newton force gives 250 micronewtons, a force 250 times larger than the largest forces produced with old style Mach effect thrusters.

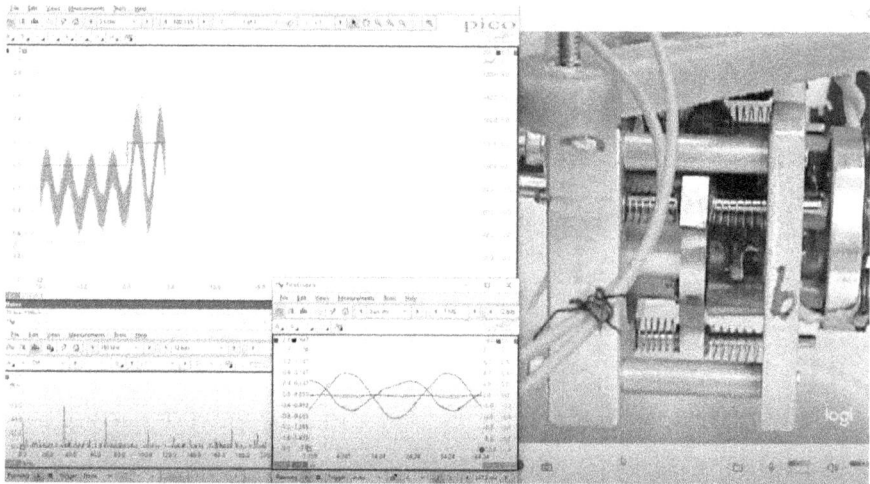

Figure 5.
A screen capture of the display screen for a run in progress. Three Picoscope displays are on the left and a movie of the device is on the right. In the upper left strip-chart recording the red and gray traces are the device and support structure positions, and the blue trace is the rectified voltage across the device. Below the strip-chart are the FFT power spectrum of the current (left, note the prominent first and second harmonics) and the waveform (right, voltage blue and current black traces) displays. The movie on the right is used to calibrate the position traces in the strip-chart so that the positions can be converted to force measurements.

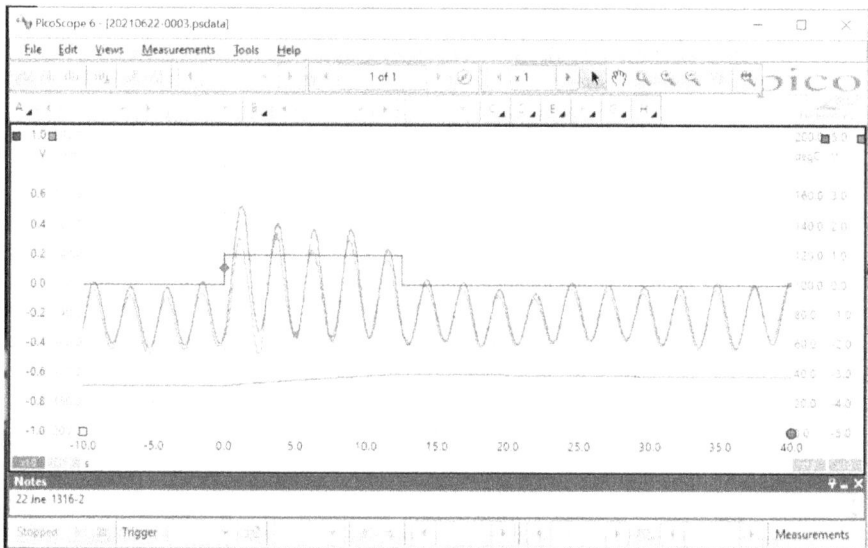

Figure 6.
The completed run. The gray trace of the support structure position has been filtered and the green temperature trace has been added.

Figure 6 (and many others like it) shows that a real force acts on the device and support structure when the device is excited. And the force continues to act as long as the device is excited. The obvious question is: can the real force responsible for the pendulum deflection in **Figure 6** be attributed to some mundane cause? The leading candidate, Newtonian vibrational artifacts [13], has already been excluded as it cannot produce a steady force and the device and support structure do not move as required by this hypothesis. The only other possibilities are coupling to the ambient air in proximity to the device and electromagnetic interactions arising from the currents and voltages present. Ambient air was excluded while working on the torsion balance before transitioning to the cantilever and then the pendulum. The operation of these devices is unaffected by operation in air at atmospheric pressure or in soft vacua of 10 milliTorr or so. Electromagnetic effects are excluded by replacing the device with a "dummy" capacitor with capacitance roughly equal to that of the devices, but without the electromechanical properties of the PZT stacks. Runs with the dummy capacitor show no signs whatsoever of any pendulum activity like that in **Figure 6** (and many others).

From the practical perspective it seems reasonable to suggest that the obvious advantages of real MEGA impulse engines will make them likely features of our future. From the physics perspective, the fact that MEGA impulse engines work constitutes experimental confirmation of Einstein's insistence on "the relativity of inertia" and the gravitational induction of inertial effects.

8. Conclusions

In the matter of the role of inertia in general relativity, we find that:

- Einstein regarded spacetime as the gravitational field. In the absence of gravity, there is no spacetime, from which it follows that

- Minkowski spacetime, used in the construction of general relativity, as it assumes the absence of gravity, is not a valid general relativistic spacetime.

- Since spacetime – the gravitational field that is – is observed to be spatially flat at cosmic scale and in sufficiently small regions, the metric that obtains where spatial flatness is the fact is that of the spatially flat FLRW cosmology.

- Carl Brans' spectator matter argument led to the rejection of Einstein's argument that inertia is gravitationally induced in general relativity.

- However, while Brans' argument is correct, the inference that it excludes gravitationally induced inertia is not correct. What Brans' argument does do is require that the total, locally measured Newtonian gravitational potential be a scalar invariant, like the vacuum speed of light, to which it is related, being the square thereof.

- This relationship between the locally measured values of the vacuum speed of light and Newtonian gravitational potential are an automatic consequence of spatially flat FLRW cosmology.

- Stipulation that inertia is to be understood as gravitationally induced in general relativity can be implemented by asserting a condition on acceptable coordinates, constraining the range of acceptable solutions of Einstein's field equations, to those that return the requisite relationship between the locally measured invariant values of the vacuum speed of light and the Newtonian gravitational potential.

In the matter of experimental confirmation of the correctness of Einstein's contention that inertia be an inductive gravitational effect, we find that:

- Extended objects capable of changing their internal energies simultaneously experience changing internal energy and proper acceleration, the action of the grav/inertial field excited by the proper acceleration amplifies the rest mass fluctuation corresponding to the changing internal energy amplifies that rest mass fluctuation. Such effects are called "Mach effects" given their dependence on inertial forces of cosmic gravitational origin first adumbrated by Ernst Mach.

- Mach effects of sufficient magnitude can be utilized for propulsion by adding a synchronous mechanical oscillation at the frequency of the Mach effect fluctuation, making a Mach effect gravity assist (MEGA) impulse engine.

- Prototype MEGA impulse engines can produce thrusts of hundreds of micronewtons and more. Straight-forward tests can eliminate mundane effects that might produce false positive results that might account for observed thrusts.

- The leading candidate for a false positive explanation of observed forces is so-called "Newtonian vibrational artifacts" induced in the device by the vibration of the lead-zirconium-titanate crystal stack in the engine.

- Using a pendulum for force detection with the current realization of the MEGA impulse engine where the engine is mounted on its support structure with rods and linear ball bushings enables a simple test of the vibrational artifact hypothesis. Conservation of momentum dictates that vibrational artifacts cannot produce a steady deflection of the pendulum. A real Mach effect force will produce a steady deflection of the pendulum.

- Vibrational artifacts, by the conservation of momentum, cause the device and its support structure to move initially in opposite directions with the system subsequently moving about zero deflection. Real force causes the device and support structure to move together with subsequent motion about a time-averaged net deflection while power is applied.

- **Figure 6** shows beyond reasonable doubt that since the device and support structure move together, a real force of about 200 micronewtons was generated by the MEGA impulse engine being tested on a pendulum that, by exclusion of mundane effects, was generated by the Mach effect.

- Further work to implement this technology is warranted.

Acknowledgements

I am indebted to Hal Fearn, Daniel Kennefick, José Rodal and Nader Inan who participated in Zoom calls every few weeks for the past several years where the nature of inertia in general relativity was the topic of focus. At times, the calls were supported by a NASA Innovative Advanced Concepts Phase II grant. The experimental work reported here was done in collaboration with Hal Fearn, Michelle Broyles, David Jenkins and Paul March. I am also indebted to Gary and Anne Hudson, and Robin Snelson of the Space Studies Institute, and the Department of Physics at CSUF for ongoing support for many years.

Author details

James F. Woodward[1,2]

1 Space Studies Institute, USA

2 Department of Physics, California State University Fullerton, Fullerton California, USA

*Address all correspondence to: jwoodward@fullerton.edu

IntechOpen

References

[1] Einstein, A., "Is There a Gravitational Effect Which is Analogous to Electrodynamic Induction?", *Vierteljahrsschrift für gerichliche Medizin und öffentliches Sanitätswesen*. 1912;**44**: 37-40.

[2] Einstein, A. *The Meaning of Relativity*, Princeton Univ. Press, Princeton (1955), 5th ed.

[3] Einstein, A. "Concerning the Aether", *Verhandlungen der Schweizerischen Naturforschenden Gesellschaft*, 1924;**105**:2, 85 – 93.

[4] Brans, C., "Absence of Inertial Induction in General Relativity." *Phys. Rev. Lett.*;1977;**39**;856 – 85.

[5] Ciufolini, I. and Wheeler, J.A., *Gravitation and Inertia* (Princeton, 1995)

[6] Sciama, D., "On the Origin of Inertia", *M.N.R.A.S.*; 1953;**34**, 34 – 42.

[7] Harris, E., "Analogy between general relativity and electromagnetism for slowly moving particles in weak gravitational fields", *Am. J. Phys.***59** 421 – 425 (1991).

[8] Pfister, H. and King, M., *Inertia and Gravitation*, (Springer, New York, 2015),

[9] Bernstein,J., *An Introduction to Cosmology*, (Prentice Hall, Englewood Cliffs,, N.J., 1995)

[10] Braginski, V., Caves, C. and Thorne, K., "Laboratory Experiments to Test Relativistic Gravity", 1977, *Phys. Rev .D1* **15**, 2047 – 2068, especially section 3.

[11] Misner, C.W,, Thorne, K.S. and Wheeler, J.A., *Gravitation*, (Freeman, San Francisco, 1973).

[12] Woodward, J.F., "Keeping the Dream Alive: Is Propellant-less Propulsion Possible?", in: *Propulsion: New Perspectives and Applications*, (IntechOpen, London, 2021).

[13] Tajmar, M., *et al.*, "Experimental Investigation of Mach Effect thrusters on a torsion balance", Acta Astranautica **182** (2021) 531 – 546,

Chapter 3

Temporal (t > 0) Space and Gravitational Waves

Francis T.S. Yu

Abstract

I will begin with the nature of our temporal (t > 0) universe, since without temporal space there would be no gravitation force because gravitational field cannot be created within an empty space. When we are dealing with physical realizability of science, Einstein's relativity theories cannot be ignored since relativistic mechanics is dealing with very large objects. Nevertheless I will show that huge gravitational waves can be created by a gigantic mass annihilation only within a temporal (t > 0) space. Since gravitational energy has never been consider as a significant component within big bang creation, I will show it is a key component to ignite the big bang explosion, contrary to commonly believed that big bang explosion was ignited by time. I will show a huge gravitation energy reservoir induced by a gigantic mass had had been created over time well before the big bang started. Since the assumed singularity mass within a temporal (t > 0) had had gotten heavier and heavier similar to a gigantic black hole that continuingly swallows up huge chunk of substances within the space. From which we see that it is the gravitational force that triggers the thermo-nuclei big bang creation, instead ignited by time as postulated. Aside the thermo-nuclei creation, it had a gigantic gravitational wave release as mass annihilates rapidly by big bang explosion. From which we see that it is the induced gravitational reservoir changes with time, but not the induced gravity changes (i.e., curves) time–space. In other words if there has no temporal (t > 0) space then there will be no gravitational waves.

Keywords: Gravitational Energy, Gravitational Waves, Gravitational Force, Einstein Energy Equation, Big Bang Creation, Curving Time–Space, Temporal Space, Timeless Space

1. Introduction

One of the essences of Einstein's general theory of relativity is curving the space–time [1]; from which as John Wheeler had said that as I quote "Space-time tells matter how to move; matter tells space-time how to curve". However as I see it; it is time tells space how to curve but "not" space tells time how to curve. Nevertheless Einstein's general theory of relativity was developed based on a Minkowski type space–time continuum where time is treated as an "independent" variable (i.e., an independent dimension) [2]. However from temporal (t > 0) universe standpoint, time and space are coexisted in which time is a "dependent" forward variable moving at a "constant" speed. In other words within our temporal universe, time curves time–space, but time–space "cannot" curve the pace of time.

Since it is impossible to create a magnetic field within an empty space that normally assumed it could. But we will show it is a temporal (t > 0) space, instead of an empty space that normally assumed [3], that an assumed gigantic mass was situated. For which the mass is capable for continuingly attracting substances to build up a super-gigantic mass, such that a huge gravitational field induced by the mass can be established overtime. From which we see that it was the huge convergent gravitational force that ignited the thermo-nuclei big bang explosion created our universe, but not by time. In other words it was big bang explodes with time, but not time ignites the big bang.

2. Nature of temporal (t > O) universe

As we accepted subspace and time are coexisted within our temporal (t > 0) universe [4, 5], time has to be real, and it cannot be virtual since we are physically real. And every physical existence within our universe is real. The reason some scientists believed time is virtual or illusion is that; it has no mass, no weight, no coordinate, no origin, and it cannot be detected or even be seen. Yet time is an everlasting existed real variable within our known universe. Without time there would be no physical matter, no physical space, and no life. The fact is that every physical matter is coexisted with time which including our universe. Therefore, when one is dealing with science, time is one of the most enigmatic variables that ever presence and cannot be simply ignored. Strictly speaking, all the laws of science as well every physical substance cannot be existed without the existence with time. For which we see that; time cannot be an independent dimension or an illusion. In other words, if time is an illusion, then time will be independently existed from physical reality or from our universe. And this is precisely that many scientists have treated time as an independent variable such as Minkowski's space [2], for which we see that Einstein 's space–time can curve time. However if matter can curve time–space, then we can change the speed of time. But as I see that it is our universe exists with time, it is "not" our universe changes time.

Since time is a constantly moving dependent variable at a constant pace, for scientific presentation we usually use numeric symbols to represent time otherwise it would be very difficult to facilitate and to understand the nature of time. For convenience we had divided time into past (i.e., t < 0), present (i.e., t = 0), and future (i.e., t > 0) domains to represent time, as exemplified in **Figure 1**.

From which we see that our universe changes with time; for example present moment at t = 0 moves immediately forward to become the next present moment (t + Δt). In other words the present moment t = 0 becomes the moment of past. Once the present moment (t = 0) moves forward a section of Δt → 0, no matter how small it is, it is impossible to return back, since our universe changes with time. From which we also see that it is impossible to move the current moment (t = 0), no matter how small Δt is, ahead or behind the pace of time. Nevertheless this diagram exemplifies our temporal (t > 0) universe changes with time, since our universe is a stochastic dynamic temporal (t > 0) space [4, 5]. Of which we see that it is impossible to travel backward or ahead the pace of time.

Since past time domain (i.e., t < 0) represents the moment of certainty events (e.g., past universes), they were the past memories (i.e., information) but without physical substance in it and no time. Which is similar as viewing a backward video clip, if we move time backwardly (t < 0), we see that past consequences (i.e., past universes) changes with time (e.g., t = − t$_n$) as a backward movie clip. In view of Einstein's general theory of relativity as I quote; matter (i.e., time–space) curves (or changes) time–space. And this is precisely the section of past time (i.e., t < 0)

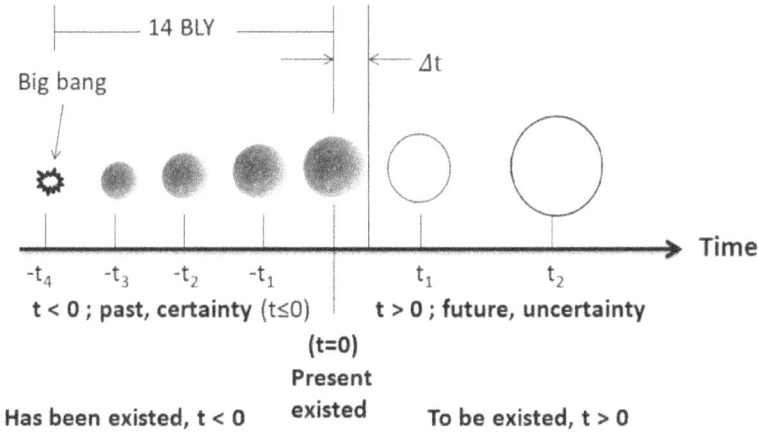

Figure 1.
Shows that our temporal (t > 0) universe changes naturally with time, in which it shows the age of our universe is about 14 billion light years old. The past time domain (t < 0) represents a set of certainty virtual events (i.e., past universes), the future time domain (t > 0) represents a physically realizable domain of uncertainty. And the instantaneous present moment (t = 0) is the only moment of absolute physical certainty. Yet we see that present moment is instantaneously moved forwardly to become the next new present moment [(i.e., t = 0 + Δt) where Δt → 0], to next absolute certainty moment.

domain that Einstein used to derive his general theory, by which his theory had have treated time as an independent variable. This is precisely why general theory is a deterministic theory instead indeterministic. Which is similar to most of the classical sciences are deterministic, yet science is supposed to be non-deterministic or approximated.

Nevertheless within our temporal (t > 0) universe, time is a dependent or interdependent variable with respect to the subspace since space and time are coexisted. In which we see that future events (i.e., t > 0 domain) are non-deterministic consequences with degree of uncertainties. And this is the positive time [or temporal (t > 0)] domain that Einstein general theory may not apply within t > 0 domain, since subspaces are not deterministic (i.e., our universe changes with time). Nevertheless the implication of temporal (t > 0) is that physical realizable events exist if and only if within positive time domain, by which the instantaneous t = 0 can only be approach but never be able to attain (i.e., t ⟶ 0), even assumed we have all the energy (i.e., ΔE) to spend.

To further epitomize the nature of our temporal (t > 0) universe, I have come up with a composite diagram as depicted in **Figure 2**, which shows that our universe started from a big bang creation, although time has been existed well before the creation. Since the past certainty consequences (i.e., memory spaces) were happened at specific time within the negative time domain (i.e., t < 0), we see that every specific past time event had have been determined with respect to a specific past certainty subspace. From which we see that time can be treated as an independent variable with respect to the past certainty consequences in the pass-time domain (t < 0) as from mathematical standpoint. But from physical reality standpoint, time is no longer existed within the past time (t < 0) domain. And this is precisely why time can be treated as an independent variable from mathematical analysis to predict what would happen at a distant future, but with some degrees of uncertainty since physical substance or subspace changes naturally with time.

This is precisely what classical laws and principles had done to science, using past deterministic certainties to predict the future. Since prediction is supposed to be non-deterministic (i.e., uncertainty), yet all the predicted solutions were

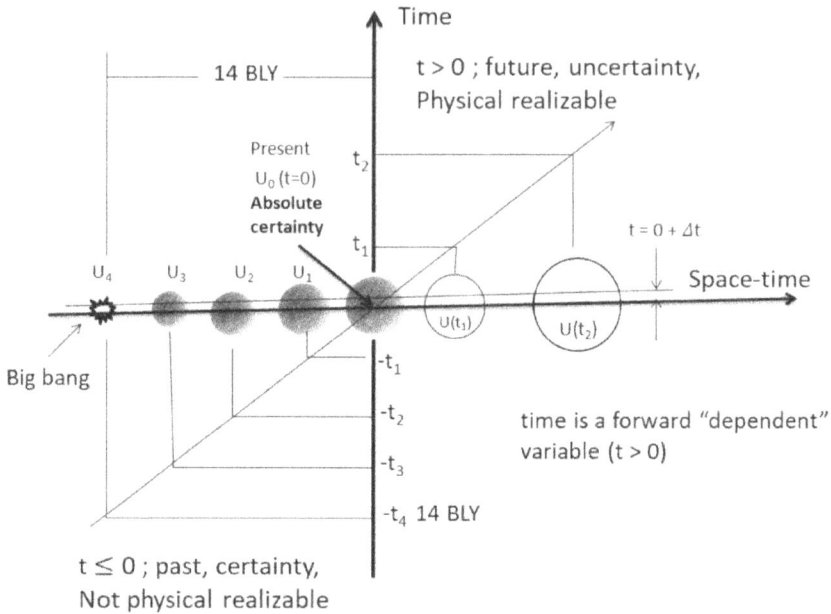

Figure 2.
Shows a composited temporal (t > 0) time–space diagram to epitomize the nature of our temporal universe. BLY is billion of light years. In which instant-present moment t = 0 is the only moment of absolute physically certainty of our universe. Past time domain (t < 0) shows past certainty universes but without time and no physical substance. And future time domain (t > 0) represents a physically realizable domain that changes with time.

maintained deterministic. it is because deterministic analysis produces deterministic solution. In other words all physically realizable solutions strictly speaking should be temporal (t > 0) solution. From which we see that all the laws, principles, theories as well paradoxes were developed from the deterministic past certainties to predict the future uncertainty consequences, but in reality, those laws and principles should not deterministic instead of deterministic. From which we see that Einstein's general theory of relativity cannot be the exception [2]. But using past deterministic to predict the future consequence is likely be deterministic, that contradicts with our temporal (t > 0) universe, which is a non-deterministic universe (or subspace) constantly changes with time.

Although using past certainties to predict future outcome is a reasonable method that had have been using for centuries, but it is physically wrong if we treated time as an independent variable within our temporal (t > 0) universe. From which we see that irrational, weird, and fictitious solutions emerged, which had had already been dominating the world-wide scientific conspiracy. This includes Schrödinger 's fundamental principle of superposition [6], Einstein's special and general relativity theories [2], Hawking's space–time [7] and others. Since they were all developed from the past certainties to predict the deterministic future, but future prediction physically cannot be deterministic or certainty.

Nevertheless the section of time Δt shown in **Figure 2** represents an incremental moment after the instant t = 0 moves to a new t = 0 + Δt, where Δt can be as small as we wish (i.e., $\Delta t \longrightarrow 0$). Yet we will never be able to squeeze it to zero (i.e., $\Delta t = 0$) and this is the section of time that cannot be delay or moved ahead the pace of time (i.e., t < 0 + Δt or t > 0 + Δt) or even stop. From which we see the aspect for time traveling either ahead or behind the pace of time is inconceivable, since we are coexisted with time.

Moreover the present instant $t = 0$ which represents the absolute certainty moment within our temporal $(t > 0)$ universe, and this is the moment of time (i.e., $t = 0$) that divides the physical and virtual realities. From which we see that; future uncertainties are physical realizable consequences, but all the past deterministic consequences were the virtual reality since time is no longer existed within the past time domain i.e., $(t < 0)$. From this conjecture we see that any hypothetical solution obtained from the past deterministic domain is anticipated to be deterministic. Nevertheless, every aspect happens within our temporal $(t > 0)$ universe is a physical reality but non-deterministic. In other words every physical reality within the temporal $(t > 0)$ universe are uncertainties that change with time. In other words, any deterministic science within our universe temporal $(t > 0)$ is virtual as from strictly physically realizable standpoint. This is precisely the reason that classical sciences are deterministic. But this by no means that timeless $(t = 0)$ solutions are useless, the fact is that all the laws, principles and theories were developed from the past certainty regime are still the foundation of our science. From which it tells us that; science developed mostly from the past certainties were deterministic, but science within our temporal $(t > 0)$ universe are probabilistic or non-deterministic which changes with time.

Nonetheless, without the past deterministic consequences, it has no better way to determine the non-deterministic consequences. Thus we see that if temporal $(t > 0)$ constraint is imposed on the past deterministic consequences in search for future non-deterministic solution, very likely physically realizable solution would emerge. From which we see that science is not supposed to be deterministic, science is a law of approximation. In view the nature of temporal $(t > 0)$ space we see a temporal $(t > 0)$ exclusive principle as stated: Empty space and temporal $(t > 0)$ space are mutually exclusive.

Since physically realizable paradigm is depending on temporal space, it is vitally important to have a basic idea of our temporal $(t > 0)$ universe, for which we exemplify the nature our temporal $(t > 0)$ universe as depicted in **Figure 3**.

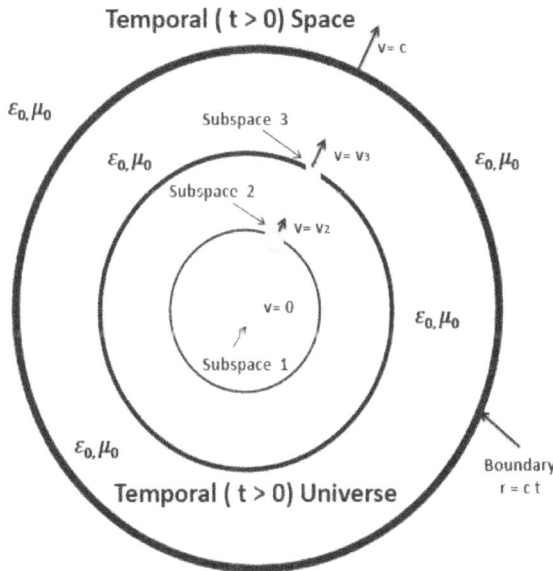

Figure 3.
Shows a simplified diagram of our temporal $(t > 0)$ universe. c is the speed of light. v is the radial velocity. In which we show that every subspace is moving radially toward the boundary of the universe, which is linearly proportional to the speed of light since light speed is the current limit.

In which we see that every subspace has the same time within the entire universe. And our universe also has the same time with the same pace as the greater temporal pace that our universe is embedded in.

3. Gravitational energy

As we have accepted the origin of our universe was started by a big bang explosion from a gigantic mass annihilation within in a temporal (t > 0) space [4, 5], instead within an empty space as normally assumed [3]. Then before the big bang started a question may be asked, what triggers the explosion? As I will show that it must be ignited by an intense convergent gravitational force, induced by a gigantic mass M(t), that triggers the thermo-nuclei explosion which is mass to energy conversion.

Since big bang creation cannot be started from an empty space, big bang creation has to be started within a temporal (t > 0) because mass M(t) is temporal (t > 0), then a question is asked, under what physical means that will ignite the big bang explosion? I assert that it must be triggered by an extreme "convergent" gravitational force induced by mass M(t) over time as depicted in **Figure 4**.

From which we see that a huge mass M(t) had had been existed within a temporal (t > 0) space well before the big bang started. Since temporal space is a non-empty space, it allows M(t) to continuingly attracting new substances into mass M(t). Then eventually a huge induced gravitational field was created as M(t) grows. In which we see that mass M(t) is able to attract more and more substances added to her mass. Eventually M(t) behaves like a giant Black Hole or it is a black hole [8] that swallows more and more substances over time. From which we see that as M(t) is getting heavier and heavier until her "storage" gravitational pressure reaches to a point that triggers the thermo-nuclei explosion of mass M(t). From which we see that it must be the induced gravitational force that triggers the big

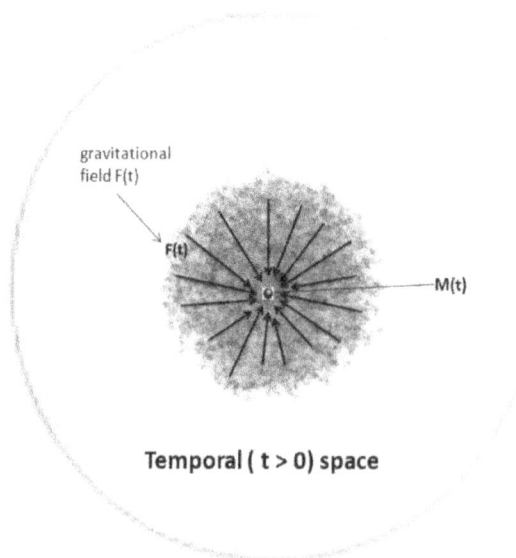

Figure 4.
Shows a well before the big bang explosion scenario. In which the dark dot represents a point-singularity approximated gigantic mass M(t), F(t) represents a huge gravitational field induced by M(t), the arrows show a set of very intense "convergent" gravitational force are applying at M(t).

bang explosion instead by "time" as most cosmologists assumed [3]. Thus it is reasonable to accept that, if mass M(t) were not embedded within a temporal (t > 0) space then there would be "no" gravitational field to create by mass M(t) and there would be "no" big bang. This is one of the many examples shows that physically realizable science comes from a physically realizable paradigm. From which we had have seen virtual and fictitious conjectures based the big bang creation within an empty space and it is hard to accept those illogical predictions as from physically realizable scientist standpoint.

4. Big bang and gravitational field

Since mass M(t) and her induced gravitational field are temporal (t > 0) substances, by which induced gravitational field "coexists" with mass M(t) as given by,

$$F(r; t) = G \frac{m\, M(t)}{r2} \tag{1}$$

from which we see that gravitational force strength F(r; t) "decreases" rapidly as inverse square law of distance r, where G is a gravitational constant and m represents an unit reference mass (i.e., points of interest) as illustrated in **Figure 5**.

With reference to the point of interest, "potential" energy for each unit m away from gigantic mass M(t) is given by [9];

$$E' = G_0\, M(t)/r \tag{2}$$

where $G_0 = G \cdot m$ is a "normalized" gravitational constant. In which it shows that gravitational energy exponentially "increases" as distance approaches to mass M(t).

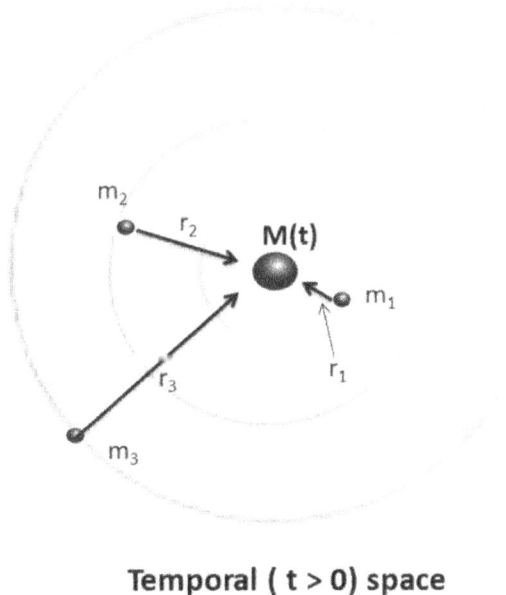

Temporal (t > 0) space

Figure 5.
Shows induced gravitational forces converge at a point-singularity approximated mass M(t). M represents a unit mass of interest. In which we see that without embedded within a temporal (t > 0) space paradigm it is impossible to create an induced gravitational field stored around mass M(t).

From which we see that as mass M(t) "reduces" rapidly with time, magnetic force attached to m (i.e., point of interest) releases quickly that causes m moves outwardly away as the induced gravitational force loses her pull. The outward force acted on each m, by Newtonian second law is approximated as given by,

$$f \approx ma \tag{3}$$

where f is an outward acting force on unit m and a is its acceleration.

$$a = \frac{G\,M(t)}{r2} \tag{4}$$

which is proportional to the inversed square laws of distance r. From **Figure 6** we see that further away from M(t) is lowering the acceleration a. While closer to M(t), acceleration of m is anticipated to be very high, as the gravitational field shrinks rapidly. This rapidly disappearing gravitational field give rise to a huge amount of energy as mass M(t) annihilates itself rapidly with time. From which we see that a gigantic gravitational energy together with a huge thermo-nuclei energy are simultaneously releasing as the big bang started.

Yet, without the thermo-nuclei mass annihilation there would be "no" such magnitude of gravitational waves that can be detected [10]. Unlike the electro-magnetic waves, gravitational waves are mostly "longitudinal" waves which dissipated quickly due to mass in motion within our temporal (t > 0) universe. As in contrast with transversal electro-magnetic wave it travels at speed of light. From which we see that it is extremely difficult to detect gravitational waves due to mass or masses in motion within our universe as can be seen as depicted in **Figure 7**.

Nevertheless the essence of preexisting temporal (t > 0) space condition is very crucial since any analytical conjecture or solution comes out from a physically

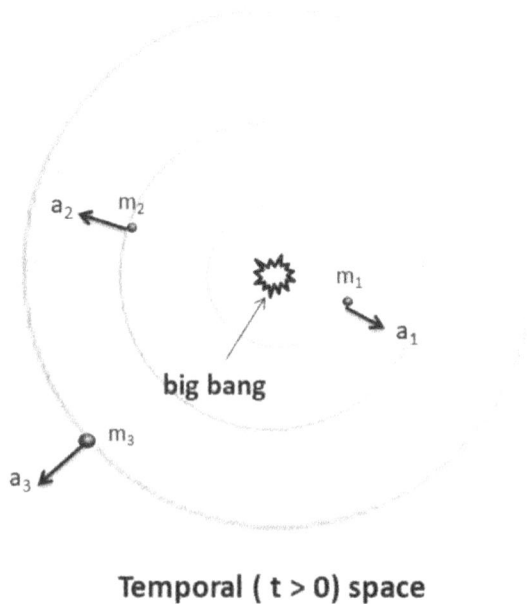

Temporal (t > 0) space

Figure 6.
Illustrates the thermo-nuclei big bang hypothesis, where the associated gravitational field releases its energy as the stored gravitational field shrinking with time rapidly. In which we see that unit m moves outwardly as gravitational field shrinks rapidly with mass M(t) annihilates.

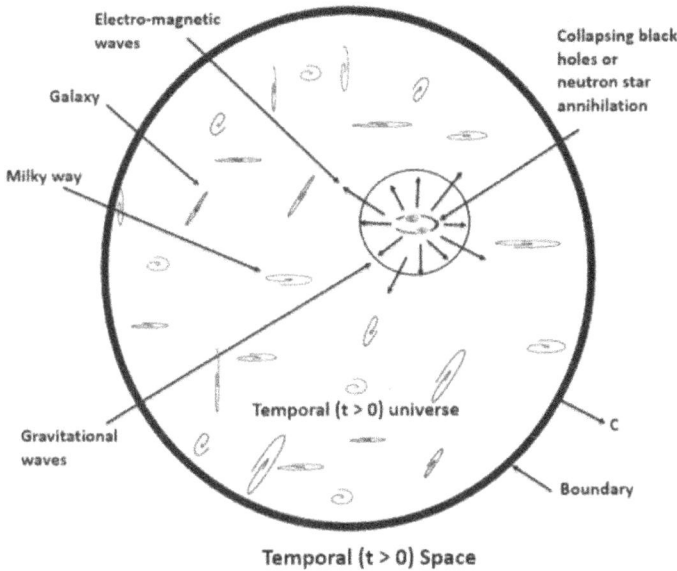

Figure 7.
Shows a scenario of possible black holes collide-annihilation or neutron star explosion. Aside the anticipated electro-magnetic energy radiation at speed of light, a huge gravitational waves releases as mases of black holes annihilation as depicted in the figure.

realizable paradigm is "likely" to be physically realizable, as in contrast with commonly used paradigm gravitational field can be created within an "empty" space. Since substance and emptiness are mutually exclusive, empty space is a "non-physically" realizable paradigm [11]. Aside the non-physically realizable issue, empty space has "no" substances for gravitational field to store. From which we see that it is a physically realizable reason to assume that big bang explosion was triggered by a huge convergent gravitational force induced by mass M(t), instead triggered by time as some cosmologists believed.

5. Nature of gravitational waves

Strictly speaking there are "two" dominant energies that associated with mass M before big bang explosion as given by,

$$E' = G_0 \, M/r \qquad (5)$$

$$E = (\tfrac{1}{2}) \, Mc^2 \qquad (6)$$

where E' represents the gravitational energy induced by mass M, and E is the thermo-nuclei energy due to mass M annihilation. Since physically realizable paradigm guarantees her solution would be physically realizable, but either Eq. (5) and Eq, (6) are not physical realizable. Firstly there are timeless (t = 0) or time independent equations, as most of the laws and principles do. Secondly there are not temporal (t > 0) equations yet since mass M does not change with time [i.e., or temporal (t > 0)]. In which we see that everything existed within a temporal (t > 0) space has to be temporal (t > 0). Thus from physical reality standpoint, the existence mass M it has to be temporal (t > 0) [i.e., M(t)]. Which means that M(t) changes naturally with time and exist within positive time domain. For which

Eq. (5) and Eq. (6) can be written in temporal (t > 0) formulas as given by, respectively,

$$E'(t) = G_0 \, M(t)/r, t > 0, \tag{7}$$

$$E(t) = (½) \, M(t) \, c^2, t > 0, \tag{8}$$

where t > 0 denotes that equation is complied with the temporal (t > 0) condition (i.e., exists within the positive time domain (t > 0). E' (t) is the gravitational equation, E(t) is the thermo-nuclei energy equation, and M(t) is a temporal (t > 0) mass.

In view of thermo-nuclei Eq. (8) one might wonder where the (1/2) factor comes from since it is different from Einstein's energy Eq. $E = Mc^2$. For which I will show in a in Section 6 Einstein's energy equation is physically significantly correct that energy and mass are equivalent, but it is "not" physically realizable within our temporal (t > 0) universe. It is because Einstein energy Eq. $E = Mc^2$ was derived from his special theory of relativity, but his special theory was developed within a non-physically realizable empty space. Since $E = Mc^2$ and $E = (½)Mc^2$ shares identical physical significance that energy and mass are equivalent, but $E = (½)Mc^2$ was based on kinetic energy standpoint where velocity of light is the current physical limit.

However, it is the induced gravitational energy E'(t) that had had never been a component included within the big bang explosion that I am concerned [3]. For which we start with the total potential energy due to induced gravitational field of Eq. (7), as referenced to point of interest "m" the overall gravitational energy induced by mass M(t) can be "approximated "by,

$$E''(t) \approx - G_0 \, (4/3\pi)(r_0)^2, t > 0 \tag{9}$$

this shows that total gravitational energy E''(t) decreases as mass M(t) annihilates. From which we see that a huge amount of gravitational energy releases instantly soon after M(t) annihilated. In other words an intense "divergent" gravitational shock waves releases almost simultaneously with thermo-nuclei explosion, within a newly created expanding universe as depicted in **Figure 8**.

Since Eq. (7) and Eq. (8) are not time varying equations, strictly speaking they "cannot" implement directly within the temporal (t > 0) space unless they were reconfigured into time-varying partial differential forms, as given by [4, 5],

$$\frac{\partial E''(t)}{\partial t} \approx - K \frac{\partial M(t)}{\partial t} = [\nabla \cdot S''(t)], t > 0 \tag{10}$$

$$\frac{\partial E(t)}{\partial t} \approx - \left(\frac{1}{2}\right) c^2 \frac{\partial M(t)}{\partial t} = [\nabla \cdot S(t)], t > 0 \tag{11}$$

where $K = (4/3)\pi \, G_0 \, (r_0)^2$, $\nabla \cdot$ represents a divergent operator, S''(t) and S(t) are the respective gravitational and thermo-nuclei energy vectors and (t > 0) denotes that equation is subjected to the temporal (t > 0) constraint. In other words equation only exists in the positive time domain or equivalently temporal (t > 0).

As we know that an equation is a language, a picture or even a video, from which we see that soon after the big bang explosion two divergent energies emerge from the exploding mass M(t) are illustrated in **Figure 8**, one is due to thermo-nuclei explosion and the other is from sudden releases (i.e., outward explosion) stored gravitational energy due to instantaneous mass M(t) annihilation. Although thermo-nuclei explosion is responsible mostly for the big bang creation [4, 5] for

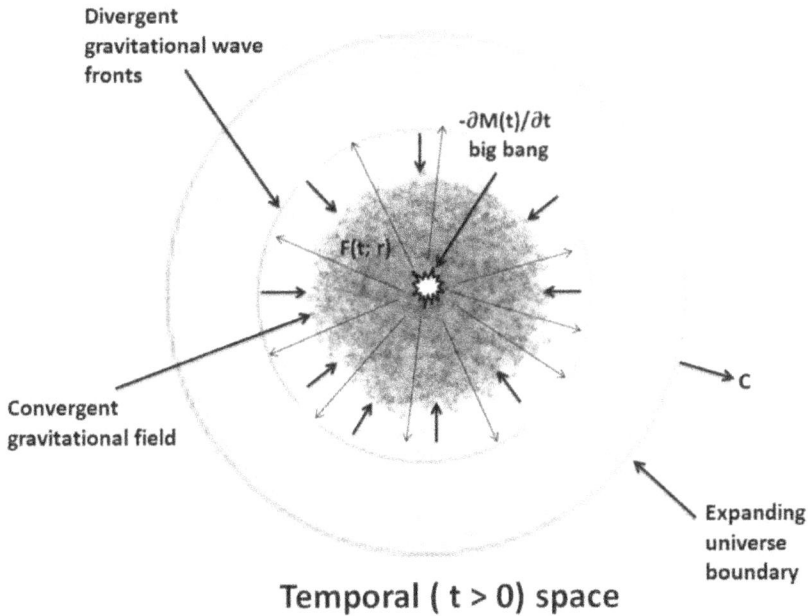

Figure 8.
Shows a composited diagram that our universe was created. The set of converged arrows represents a shrinking gravitational field. A set of outward arrows shows an outward energy explosion due to big bang. In which we also see that the boundary our universe is expanding at speed of light due to thermo-nuclei big bang explosion.

which the boundary of our universe is expanding at the speed of light, but with a surge of gravitational waves as represented by a set of arrows diverges from the big bang explosion as can be seen in the figure. From which we see that a set of convergent arrows represents the collapsing gravitational field as the mass of M(t) reduces rapidly as big bang explosion started.

Since every subspace within our universe is created by an amount of energy ΔE and a section of time Δt, we see that it is the "necessary cost" for space creation, which includes our universe herself. For instance mass to energy conversion can be written in partial differential form as given by,

$$\frac{\partial E(t)}{\partial t} \approx -\left(\frac{1}{2}\right) c^2 \frac{\partial M(t)}{\partial t}, t > 0 \tag{12}$$

In which we have ignored the stored gravitational energy due to mass M(t), since thermo-nuclei energy is much greater than the induced gravitational energy from mass M(t) [i.e., E(t) > > E"(t)], where t > 0 denoted that equation is subjected to temporal (t > 0) condition or exists only in the positive time domain t > 0. By which the "total" amount of energy due to big bang explosion can be approximated by,

$$\Delta E(t) \, \Delta t \approx (\frac{1}{2}) \, M_0 \, c^2 \tag{13}$$

where M_0 represents the total mass and c is the speed of light. Since $\Delta E(t) \, \Delta t$ is equivalent to a temporal (t > 0) subspace. In this case we see that our universe changes with time [i.e., temporal (t > 0)].

For example if we let t = 0 which is at the time equals to 14 BLY (i.e., billion light years) after the big bang, the amount of energy ΔE and the section of time $\Delta t = 14$ BLY that created our universe is given as,

$$\Delta E(t = 0) \, (\Delta t = 14 \text{ BLY}) \approx (\tfrac{1}{2}) \, M_0 \, c^2 \tag{14}$$

where t = 0 represents the instant present moment, after the big bang explosion 14 BLY ago which is the moment when big bang started to explode (i.e., 14 BLY ago). Nevertheless, Eq. (13) can be written as.

$$\Delta E_{t<0} \, \Delta t_{t<0} \approx (\tfrac{1}{2}) \, M_0 \, c^2 \tag{15}$$

where t is bounded between - 14 BLY to 0 [i.e.,(− 14 BLY, 0)], and Δt increases proportionally from − 14 BLT to 0. In view of preceding equation we see that energy is conserved which is equals to the total equivalent energy of the big bang mass M_0. From which we see that the section of time Δt = 0 means that no energy releases yet from mass M_0 at exactly 14 BYL ago (i.e., t = −14 BLY). In other words our past time universes [i.e., $\Delta E_{t<0} \, \Delta t_{t<0}$] can be treated as a time-independent universe from mathematical standpoint since time and physical substance are no longer there. And this is the past-time universes (or subspaces) were deterministic (or certainty) time-spaces which we normally used to predict the future universe (or subspace). And this is precisely why all our laws, principles, and theories were deterministic instead of non-deterministic or uncertainty. Yet from physically realizable standpoint, future prediction is supposed to be non-deterministic and uncertain. And this exactly why Einstein's general theory is deterministic instead of non-deterministic, which violates the nature of our temporal (t > 0) universe, where future is hard to predict.

But as time moves on forwardly from the present t = 0 into the future time domain (I.e., t > 0), our universe [i.e., ΔE(t > 0)Δt] is an indeterministic or uncertainty domain, for which we have the following expression after Eq. (13),

$$\Delta E_{t>0} \, \Delta t_{t>0} \approx (\tfrac{1}{2}) \, M_0 \, c^2 \tag{16}$$

which shows our universe [i.e. $\Delta E_{t>0} \, \Delta t_{t>0}$] changes with time and it does not change time. From which it is a mistake to treat our temporal (t > 0) universe as a deterministic universe, as Einstein's general theory did. From which we have seen that scores of fantasy time-traveling scenarios back to the past or to the future emerged.

Yet, it remains to be answered when the section of time Δt approaches to infinitely large (i.e., Δt → ∞)? Or is our temporal (t > 0) universe having a life? As we accepted our temporal (t > 0) universe, then it would be the end of physical realizability as Δt → ∞ that must be the end of our universe. But in view of energy conservation we see that when Δt → ∞ then ΔE → 0, we should have a finite energy preserved within a huge cosmological subspace within a vast temporal (t > 0) space that our universe was created as given by,

$$(\Delta t \rightarrow \infty) \cdot (\Delta E \rightarrow 0) = (\tfrac{1}{2}) \, M_0 \, c^2 \tag{17}$$

And this is the end of our universe at t → ∞ at point of infinity, since time within the greater temporal (t > 0) space that had had supported the big bang creation of our universe has no beginning and has no end. But our universe has a beginning, but it has no end in time and in space. Similar to a wave created on a still water pond, it has the beginning, but it has no end from strictly speaking viewpoint.

Yet every subspace within our temporal (t > 0) universe, no matter how small it has a lowest limit by Planck constant. In which we see that the lowest limit for a tiniest particle within our temporal universe even at point of infinity (i.e., t → ∞) Δt ΔE is still within the quantum limit as from current knowledge of science is given by,

$$\Delta t \, \Delta E = h \tag{18}$$

where h is the Planck's constant.

Nevertheless as from macroscopic standpoint every subspace no matter how big it is, it is currently limited by.

$$\Delta t \, \Delta E = (\tfrac{1}{2}) \, M \, c^2 \tag{19}$$

where M is the mass.

Nonetheless, every subspace, as well our universe, changes with time. But our universe and her subspaces "cannot" change the speed of time since time and subspace (i.e., substance) are coexisted. Thus every subspace within our universe has the "same" time speed. Since the universe as a whole run at "the same" pace of time. In which we see that if any subspace has a "different" pace of time it "cannot" exist within our universe, that includes the timeless (t = 0) subspace. The fact is that; those timeless (t = 0) and time-independent subspaces are virtual and "non-physically" realizable subspaces. For which is "incorrect" to assume those virtual and non-physically realizable spaces as "inaccessible" subspaces within our universe as some scientists do.

6. Non-realizable relativistic theories

Since there are two pillars of modern physics one is dealing with very small particles of Schrödinger's quantum mechanics [6] and the other is dealing very large object of Einstein's relativistic theories [2], yet both of them are timeless (t = 0) or time independent principles since both of them were developed within a non-physically realizable paradigm. Firstly we see that Einstein's special theory of relativity was developed on an empty space paradigm as depicted in **Figure 9**.

In which we see that it is not a physically realizable paradigm by virtue of temporal exclusive principle. Nevertheless Einstein's special theory can be developed with Pythagoras theorem as given by,

$$\Delta t' = \Delta t / \left[1 - (v/c)^2 \right]^{1/2} \tag{20}$$

where v is the velocity of a coordinate system and c is the speed of light. Since within empty space paradigm it has no time and has no direction, Einstein's special theory of Eq. (20) shows no sign of relativistic direction. Although the implication is relative-directional similar to the kinetic energy equation it has no sign of direction, but the equation implies that the energy is on the same direction of the velocity vector v. From which we see that scientists have frequently treated special theory as a relativistic-directional independent, which is due to the empty space paradigm. The question is that why we made those trivial mistakes? The answer is that, since scientists are mathematicians, they can implant virtual time on a piece of paper as they wish. But not knowing the background of that piece had have been assumed as an empty subspace for centuries.

On the other hand, if Einstein's special theory is developed within a temporal (t > 0) subspace as depicted in **Figure 10**. For example, derivation can start at time t = t₁ with a light emitter of S, where t is the time of the background temporal (t > 0) space. With reference to the diagram, we see that it will take a section of time Δt (i.e., t = t₁ + Δt) for beam 1 to reach position 1, which is a subsection within $\Delta t'$ (i.e., $\Delta t < \Delta t'$) for light beam 2 before reaches position 2. Since v·Δt is a

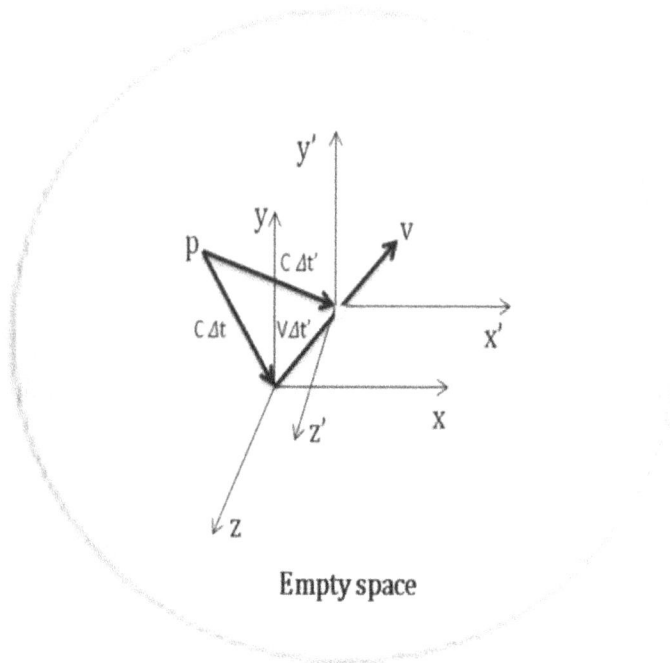

Figure 9.
Shows where Einstein's special theory of relativistic mechanics was developed from an empty space paradigm. In which we see a coordinate system (X' Y' Z') is translating at a constant speed with respect to a stationary coordinate system (X, Y, Z).

sub-distance of $v \cdot \Delta t'$ before the moving particle reaches position 2, it will take beam 2 an additional section of $c \, \Delta t'' = c \, (\Delta t' - \Delta t)$ to reach position 2 simultaneously when the particle arrives. Therefore we see that the duration at static position 1 is actually $\Delta t' = \Delta t + \Delta t''$, instead of just Δt as shown in the special theory of relativity [i.e., Eq. (20)], from which we see that the moving particle has "no" section of time-gain relative to the static position 1, since time at position 1 and 2 are "the same" $(t = t_2 = t_1 + \Delta t')$ when moving particle reaches position 2. In which the duration at position 1 is actually $\Delta t' = \Delta t + \Delta t''$, instead of Δt as shown in the special theory of relativity. Thus we see that Einstein's special theory of relativity fails to exist within our temporal $(t > 0)$ universe. In other words Einstein's special theory of relativity is a timeless $(t = 0)$ theory which is only existed within an empty space, which has no time and no space. From which we see that it is the background of that piece of paper inadvertently that had have treated it as an empty timeless $(t = 0)$ space.

Nevertheless what is the physical significant of Einstein's special theory of relativistic to what? In view of the temporal $(t > 0)$ paradigm of **Figure 10**, we see that it is the relativistic theory of distance as given by,

$$d_r = (c-v) \cdot (\Delta t' - \Delta t'') = (c-v)\Delta t \tag{21}$$

where d_r is a relativistic distance between position S of the light source and position 1 of a moving particle both simultaneous reach position 2. From which we see that light beam has traveled a extra distance of $(c - v) \, \Delta t$ more than the particle traveled. Thus we see that Einstein's special theory of relativity is relative to distance within our temporal $(t > 0)$ subspace, instead of relative to time since we cannot change time. That means that particle and the light beam arrived position 2

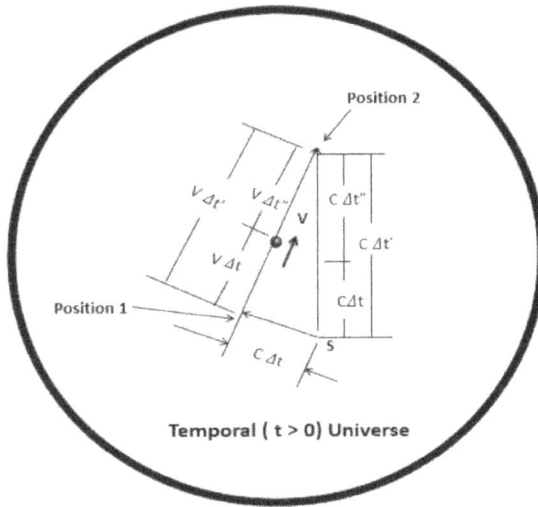

Figure 10.
Shows the same relativistic mechanics model is embedded within a temporal (t > 0) subspace. S is the light source and P is a particle in motion at a constant velocity of v, c is the velocity of light.

at the same time which is the same time at position 1, at position 2, at position S and the same time at everywhere within our universe. In which we see it has no time-gain or time-loss of the traveling particle.

Nevertheless when velocity of the moving particle approaches the speed of light (i.e., $v \to c$), we have a relative distance $d_r \to 0$. This is by no meant that time is running behind or ahead the pace of time. For which we see that it is the speed of light travels with time, and it is not the speed of light changes the pace of time.

Similarly relativistic distance of preceding equation can also be applied for relative velocity of two moving particles. For example two particles are moving at the same direction at different speeds v_1 and v_2, respectively. In view of Einstein's special theory is not a physically realizable theory within our temporal (t > 0) universe, and it is also incorrectly had have interpreted as directional-independent, as can be seen from Eq. (20). It is however should correctly treated special theory as a directional sensitive theory because of particle's velocity vector. From which we see that the relativistic distance between two particles on the "same direction" can be shown as,

$$d_r = (v_1 - v_2)(\Delta t' - \Delta t'') = (v_1 - v_2)\Delta t \qquad (22)$$

Again we have seen that Einstein's special theory is a relativistic velocity equation instead a relativistic time theory.

Equivalently Einstein relativistic mass equation can be derived from his special theory as given by,

$$M = M_0 \left(1 - v^2/c^2\right)^{-1/2} \qquad (23)$$

where M is the effective mass (or mass in motion), M_o is the rest mass, v is the velocity of the moving M and c is the speed of light. In other words, the effective mass of a moving particle increases at the same amount with respect to the relativistic time window (i.e., time dilation $\Delta t'$) increases. Nevertheless as we had shown in preceding Einstein's special theory is not a physically realizable theory within our temporal (t > 0) universe, then relativistic mass equation is also not a physically realizable equation. But one of the famous energy Eq. $E = mc^2$ was derived based on the special theory. Then the legitimacy within our temporal (t > 0) universe is in question. Since $E = mc^2$ was based to Kinetic energy equation to legitimize the significant of the equation as shown by,

$$M = M_0 \left(1 + \frac{1}{2} \cdot \frac{v^2}{c^2} + terms\ of\ order\ \frac{v^4}{c^4} \right) \qquad (24)$$

By multiply the preceding equation with the velocity of light c^2 and noting the terms with the orders of v^4/c^2 are negligibly small, above equation can be approximated by,

$$M \approx M_0 + \frac{1}{2} M_0 v^2 \frac{1}{c^2} \qquad (25)$$

which can be written as,

$$(M - M_0)c^2 \approx \frac{1}{2} M_0 v^2 \qquad (26)$$

The significant of the preceding equation is that $M-M_o$ represents an increased in mass due to motion, which is the kinetic energy of the rest mass M_o. And $(M-M_o)c^2$ is the extra energy gain due to motion. Nevertheless what Einstein postulated, as I remembered, is that there must energy associated with the mass even at rest. And this was exactly what he had proposed,

$$E \approx Mc^2 \qquad (27)$$

where E represents the total energy of the mass. In which we see that Energy and mass are equivalent but there are not equaled.

Since we had shown that Einstein's special theory of relativity exists only within empty space, from which we see his energy equation cannot be legitimized within our temporal (t > 0) universe. Yet energy and mass are equivalent is a well-accepted physical reality but may not in exact form since science after all is approximated. In view of the legitimacy and Einstein's energy equation and comparison of the well accepted although empirical kinetic energy Eq. $E = (1/2)\ m\ v^2$, where v is the velocity. Since velocity of light c is the current limit of science, it is justifiable to rewrite the energy equation in following form after kinetic energy equation as given by,

$$E \approx (1/2)\ Mc^2 \qquad (28)$$

In which we see that mass and energy are equivalent, and it has the same physical significant as Einstein's energy equation although Einstein's equruion has been illegimated. In view of preceding equation we see that energy and mass can be simply traded as given by,

$$E \leftrightarrow M \qquad (29)$$

From which in princiole we can convert mass to energy or energy to mass.

Nevertheless, one of the greatest theories that Einstein had had developed must his general theory of relativity as given by [2],

$$G_{\mu\nu} + g_{\mu\nu} = \left(8\pi G/c^4\right) T_{u\nu} \tag{30}$$

where $G_{\mu\nu}$ is the Einstein tensor, $g_{\mu\nu}$ is the metric tensor, $T_{u\nu}$ is the stress-energy tensor, is the cosmological constant, G is the Newtonian constant of gravitation, and c is the speed of light.

In view of general theory it is a point-singularity approximated deterministic equation, we see that Einstein's general theory is not a physically realizable principle since science is supposed not to be deterministic. For which it is impossible to predict future with deterministic general theory. Although we can change a section of time Δt but we cannot change the pace of time or even stop time. Strictly speaking all physically realizable theory must be temporal $(t > 0)$. For which we see that degree of uncertainty increases as time moves further away from the point of absolute certainty (i.e., present instance $t = 0$). Thus we see that it is not the complexity of mathematic that Einstein had have used, it is the physically realizable paradigm that determines the physically realizable science. Nevertheless, Einstein's theory is a relativistic theory of distance but not a relativistic theory of time since we cannot change time.

7. A necessary cost

One of most important aspects within our temporal universe is that everything has a price, and it is not free. For example, every physical realizable theory takes a section of time Δt and an amount of energy ΔE (i.e., Δt, ΔE) to implement, which is a necessary cost. From which we see that ΔE is coexisting with Δt and $\Delta E(t)$ changes with time t or temporal $(t > 0)$. Since general relativistic theory of Einstein tells us that matter curves the space–time, then space is possible to curve our universe. As in contrast with our temporal $(t > 0)$ universe, although space curves with time but space cannot curve time.

Einstein's general theory tells us it is possible to curve our universe for wormhole traveling, a scenario was proposed by renounced astrophysicists [12] as depicted in **Figure 11**, where we see a curved equivalent universe is situated within our temporal $(t > 0)$ universe. From which we see that it is possibly go through a wormhole tunnel from one edge of our universe to the other edge. Instead of crossing the vast cosmological space that will take us beyond 28 billion light years of voyage at speed of light and still unable to reach it since our universe is expanding at velocity of light. Aside the fact that **Figure 11** is a non-physically realizable paradigm (i.e., by virtue of temporal exclusive principle), my question is that how long it will take to curve the universe (i.e., a section of time Δt), in which we assume that we have all the energy ΔE we need. From which we see that the necessary cost is the section of time Δt and the amount of energy (i.e., Δt, ΔE). But in reality, to make it happen we also need an amount of information ΔI or equivalently an amount of entropy ΔS that makes it sufficient, to curving a topological equivalent universe shown in the figure.

From which we see that Einstein general theory predicts the future deterministically, but from physical reality, future is supposed to be non-deterministic or uncertainty. Of which we see that Einstein's general theory is not a physically realizable principle within our temporal $(t > 0)$ universe. Nevertheless it is possible to reconfigure his general theory to be temporal $(t > 0)$, by imposing a temporal constraint on Eq. (30) as given by,

$$G_{\mu\nu} + \wedge g_{\mu\nu} = \left(8\pi G/c^4\right) T_{u\nu}, t > 0 \tag{31}$$

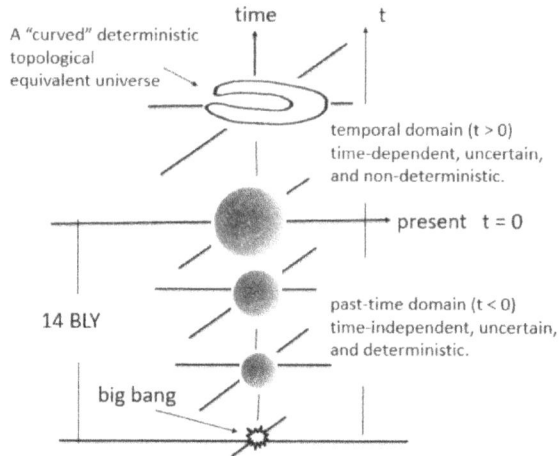

Figure 11.
Shows a non-physically realizable paradigm for curving space–time within our temporal (t > 0) universe. The curved topological equivalent space shows as a "deterministic" time–space. Since future universe is supposed to be non-deterministic or uncertain, this shows Einstein's general theory is "not" a physically realizable theory.

Where t > 0 denotes that equation is subjected by temporal constraint for which any solution comes out from this equation will be temporal (t > 0) or physically realizable.

In summary, we have seen as from Newtonian mechanics to Hamiltonian, to statistical, to wave mechanics, to relativistic and quantum mechanics are timeless (t = 0) mechanics. Although those timeless (t = 0) mechanics paved the way to our modern science, but the basic empty space paradigm had have not changed. For which it has produced a number of unthinkable virtual timeless (t = 0) solutions that causing a worldwide scientific conspiracy. Regardless of it is inadvertently or not, but it our responsibility to change it back to physically realizable science. Otherwise we will be continuingly trapping within the wonderland of timeless (t = 0) science which does not need to pay any price (i.e., Δt, ΔE). But unfortunately within our temporal (t > 0) universe everything needs a price to pay a section of time Δt and an amount of energy ΔE and it is not free.

8. Conclusion

Prior the origin of gravitational waves, I have shown the nature of our temporal (t > 0) universe. Since physical realizability of science depends on physically realizable paradigm, nature of temporal (t > 0) space paradigm supports the physical reality of science. Otherwise fictitious and virtual solution emerges which had had created a worldwide scientific conspiracy. As we are searching for gravitational waves, Einstein's relativistic theories cannot be avoided from which I had shown that his relativistic theories are not physically realizable theories since his theories were developed from an empty timeless (t = 0) space platform.

Since induced gravitational field from mass has never been a component in big bang creation, I have shown that it is a significant component for the inclusion. Prior the origin of gravitational waves, I have shown gravitational energy comes from a huge induced gravitational field by a gigantic mass within a preexistent temporal (t > 0) subspace. From which I have shown it is impossible to develop an induced gravitational within an empty space since empty space has no substance for

Temporal (t > 0) Space and Gravitational Waves
DOI: http://dx.doi.org/10.5772/intechopen.99474

gravitational energy to store. From which I had shown that without a gigantic convergent gravitational force to ignites the thermo-nuclei explosion, big band explosion was not possible to be ignited, as in contrast with commonly believed big bang explosion was ignited by time. In other words any mass annihilation will release an equivalent amount of gravitational energy associated with the mass annihilation. From which I assert that it is the gravitational field (i.e., time–space) changes with time, but not the gravitational field that changes (i.e., curves) the time–space.

Author details

Francis T.S. Yu
Emeritus Evan Pugh (University) Professor of Electrical Engineering, Penn State University, PA, USA

*Address all correspondence to: fty1@psu.edu

IntechOpen

References

[1] A. Einstein "Zur Elektrodynamik bewegter Korper" Annalen der Physik 17: 891(1905).

[2] A. Einstein, Relativity, the Special and General Theory, Dover Publishers, Inc., New York, 148-157(2001).

[3] M. Bartrusiok and V. A. Rubakov, Introduction to the Theory of the Early Universe: Hot Big Bang Theory, World Scientific Publishing, Princeton, NJ, 2011.

[4] F.T.S. Yu, "Time: The Enigma of Space", Asian Journal of Physics, Vol. 26, No.3, 143-158, 2017.

[5] F.T.S. Yu, "From Relativity to Discovery of Temporal (t > 0) Universe", *Origin of Temporal (t > 0) Universe: Correcting with Relativity, Entropy, Communication and Quantum Mechanics,* Chapter 1, CRC Press, New York, 1 -26(2019) .

[6] E.Schrödinger, An Undulatory Theory of the Mechanics of Atoms and Molecules," Phys. Rev., vol. 28, no. 6, 1049 (1926).

[7] S. Hawking and R. Penrose, The Nature of Space and Time, Princeton University Press, New Jersey (1996).

[8] S. Hawking, "Particle Creation by Black Holes", Commun. Math. Phys. 43 (3): 199-220(1975).

[9] L. D. Landau and E.M. Lifshitz, *The Classical Theory of Field,* Pergamon Press, Oxford (1951).

[10] M. Maggiore, *Gravitational Waves Volume 1, Theory and experiment.* Oxford University Press, Oxford (2007).

[11] F. T. S. Yu, "What is "Wrong" with Current Theoretical Physicists?", Advances in Quantum Communication and Information , Edited by F. Bulnes, V. N. Stavrou, O. Morozov and A. V. Bourdine, Chapter 9, p 123-143, IntechOpen, London (2020).

[12] M. S. Morris and K. S. Thorne "Wormholes in spacetime and their use for interstellar travel: A tool for teaching general relativity" Am. J. of Physics, 56 (5), 395-412(1988).

Chapter 4

Astrodynamics in Photogravitational Field of the Sun: Space Flights with a Solar Sail

Elena Polyakhova and Vladimir Korolev

Abstract

Mathematical models of the controlled motion of a spacecraft with a solar sail and the possibilities are considered, taking into account the translational motion in the gravitational field, the forces of light pressure, and rotational motion relative to the center of mass. The structure of possible problems of photogravitational celestial mechanics is proposed. To control movement, it is possible to change the size, shape, surface properties, and orientation of the elements of the sail system in relation to the flow of sunlight. The equations of motion can be presented on the basis of the problem of motion in a photogravitational field, taking into account the action of other disturbing forces. When studying the orbits of motion in the vicinity of the Earth, one should use a more general model of the photogravitational field of the restricted three-body problem. In this case, the gravitational action of the Sun and the Earth is supplemented by the field of forces of light pressure, which makes it possible to simulate real problems of dynamics.

Keywords: light pressure, photogravitational field, space flight, solar sail, spacecraft

1. Introduction

Even in ancient times, scientists, studying the relative movements of various bodies, have always tried to determine the reasons, principles and patterns that determine these movements. The combination of modern methods of mathematics, physics and mechanics makes it possible to form mathematical models of the interaction of material objects and predict possible new states based on the observed initial values of parameters and quality characteristics for the selected system of bodies.

The gravitational field for determining the forces of mutual attraction was originally considered as the field of gravity on the surface of the Earth. The famous experiments of Galileo Galilei (1564–1642) made it possible to determine the magnitude of acceleration and the direction of free fall, which is considered the same for any falling body. This defines a uniform gravitational field, which gives the simplest version of a mathematical model of gravity. The appearance of the basic laws of Johannes Kepler (1571–1630) after processing the results of many years of astronomical observations by the Danish astronomer Tycho Brahe (1546–1601) makes it possible to describe the motions of the planets of the solar system and other bodies.

As a result of the publication of "Mathematical Foundations of Natural Philosophy" by Isaac Newton (1642–1727), a substantiation of the law of universal gravitation appeared, which led to the creation of new models of gravitational forces and new possible solutions to the problems of celestial mechanics [1].

Theoretical astronomy studies various variants of the motion of a selected system of bodies: stars and planets, asteroids or comets. A special class of problems appears when the forces of attraction are supplemented by the forces of light pressure [1–7], acting on the surface of bodies. The history of the emergence of new directions of science in the form of photogravitational celestial mechanics (PhCM) is determined by remarkable scientists: Kepler, Bessel, Maxwell, Bredikhin, Lebedev and some authors.

The principle of movement in space under a solar sail is based on the effect of light pressure, which Johannes Kepler guessed about when observing the movement of comets. Kepler was the first to suggest that cometary tails are a stream of particles thrown by the action of light away from the Sun as the comet approaches the Sun (as was noticed by ancient Chinese astronomers, who did not try to explain this, however). Back in 1619, he wrote: *"Dirty matter clumps together, forming the head of a comet. The sun's rays, falling on it and penetrating through its thickness, again transform it into the thinnest substance of the ether and, leaving it, form a strip of light on the other side, which we call a comet's tail. Thus, a comet, throwing out a tail from itself, thereby destroys itself and is destroyed "*[2]. Kepler was the first to formulate the basic laws of planetary motion, and also realized and pointed out the essential role of solar radiation in the evolution of bodies in the solar system, in particular for comets.

In 1836, Bessel published a work on the motion of cometary particles under the influence of the Sun's gravitational force and its repulsive force, which (as assumed) changes like the gravity force inversely proportional to the square of the distance from the center of the Sun to the point of observation. Later, it was noted that the magnitude of the force of light pressure depends on the shape and surface area of the body, as well as on the reflection coefficient and location relative to the flow of sunlight, all other things being equal.

The mechanical theory of cometary forms was further developed in the works of the Moscow astronomer academician F.A. Bredikhin. Improving Bessel's theory, he corrected a number of inaccuracies in it and found the law of motion of a particle under various physical conditions.

New possibilities in explaining the nature of these forces appeared after it was predicted and then experimentally confirmed the effect of light (as a particular manifestation of an electromagnetic field) on material bodies. D.K. Maxwell in the middle of the nineteenth century, developing the theory and equations of the action of the electromagnetic field of forces, showed that light must produce pressure on the surface placed in the path of the light flux. Maxwell in 1873 succeeded in theoretically predicting the magnitude of the light repulsion force and substantiating the dynamic essence of light pressure as a physical effect. Experiments that confirmed Maxwell's prediction were carried out by the Russian physicist Pyotr Nikolaevich Lebedev (1866–1912).

Prominent Russian experimental physicist P.N. Lebedev, the founder of the first Russian scientific physics school, who made a significant contribution to the development of astrophysics, primarily by his virtuoso experimental proof and measurement of light pressure on solids and gases (1899; 1910). He worked in close contact with the largest Russian astrophysicist F.A. Bredikhin in the study of comets. It has been proven that comet tails are a real formation from matter flowing from the comet's nucleus. He began to experimentally prove and measure the pressure of light on solids in 1897. In 1899, for the first time in the history of science, he experimentally discovered and measured the pressure of light on solids and reported this in Lausanne (1899), and later more fully at the International Congress of Physicists in Paris (1900). The results were published in 1901. in Leipzig in "Annalen der Physik" in the article "Experimental investigation of light pressure."

It is important to note that the most consistent and logically grounded explanation of the phenomenon of light pressure was given only within the framework of quantum theory, on the basis of corpuscular concepts of electromagnetic radiation. It is here that it is possible to obtain an explicit analytical expression for the force of light pressure acting on a particle of a comet's tail, relying on a number of basic principles and relations of quantum theory.

2. Flying in space under a solar sail

The principle of movement in space under a solar sail is based on the effect of light pressure. Friedrich Arturovich Tsander (or F.A. Zander) (1887–1933), a Soviet scientist and inventor, one of the pioneers of rocketry, determines the emergence and development of ideas for space navigation under a solar sail. F.A. Tsander was one of the creators of the first Soviet liquid-propellant rocket and the author of the first technical design for a solar-sail spacecraft. The development of his is an engineering project for a space flight with a low-thrust engine in the form of a metal mirror. The idea of such spacefaring was first expressed by him in 1910–1912 as having a scientific and engineering sense, whereas previously it appeared in only a few works of science fiction.

In 1920, F.A. Tsander and K.E. Tsiolkovsky (1857–1935) discussed the possibility that a very thin flat sheet, illuminated by sunlight, could reach high speeds in space. As for the question of whether it is possible to use the property of photons for cosmic motion, they answered in the affirmative. Tsander was the first who not only expressed the idea, substantiating its scientific reliability and technical feasibility of its implementation, but also embodied this in 1924 into a calculated engineering design of a spacecraft with a reflecting mirror.

In 1921, a report on this project was presented by Tsander at a conference of inventors, and in 1924, it was revised and published in the journal "Technics and Life" under the title "Flights to other planets." In the same article, Tsander expressed ideas about the benefits of using ramjet engines, and about the possibility of using and constructing a solar sail and transferring energy to a moving rocket.

He applied to the Committee on Inventions for a space plane that can use huge and very thin mirrors to travel in interplanetary space. The project was presented in the form of two manuscripts, which, however, remained unpublished at that time. They will see the light only in 1961. In the middle of the twentieth century, science fiction writers again returned to solar sails (e.g., in A. Clark's story "The Solar Wind"), and then engineers and scientists. Around the same time, the short term "solar sail" appeared and took root as a successful borrowing from foreign science fiction. The idea of using a solar sail as a low-thrust engine corresponds to the historically important ideas of F.A. Tsander on space flight under the influence of light pressure and makes a feasible contribution to the development of the scientist's scientific heritage. Spacecraft flights using the energy of light pressure are no longer fiction, but the reality of projects of the present time and the near future [1, 8–13].

In 1924, Tsander, based on the formulas and results of Lebedev's experiment, proposed the first engineering design of a space sailing ship with a mirror-screen sail made of the finest metal foil. However, the problem of space navigation has remained outside the field of vision of scientists for a long time and gains popularity only in the space age. Theoretical developments and design projects are resumed, and variants of spacecraft with solar sails, different in design and shape, appear: with flat (solid, round or rectangular), such as parachutes or inflatable balloons, multi-bladed, that is, split like a helicopter propeller, honeycomb type, such as complex mirror systems and so on.

The main guarantee of the sail's efficiency is a high, close to unity, reflectivity of a light film mirror, which creates a high windage of the entire structure. At the level of modern technical capabilities, the value of windage, that is, the surface-to-mass ratio, is of the order of 1000 cm^2/g, which makes it possible to achieve an acceleration of the order of 0.1 cm/s^2, which is only six times less than the acceleration in the Earth's orbit from the gravitational actions of the Sun.

Since the second half of the twentieth century, the rapid development of the sailing theme was accompanied by an intensive increase in the number of publications in Russia and abroad. Numerous articles began to be published in magazines, and monographs appeared, in whole or in part, related to solar sails. Bold projects of sailing flights to the Moon, Mars, or Halley's comet, projects of illuminating the Earth from space with the help of a mirror sail-illuminator on the Artificial Earth Satellite, and many other fascinating technical ideas were described. In the 90s. the domestic device "Znamya" has been successfully developed and implemented, but it was not yet an interplanetary flight, but a launch into a satellite orbit in the vicinity of the Earth.

The first attempt to implement the "Znamya" project and deploy a solar sail in space was successfully carried out in 1993 [2, 4, 13]. Such space sailing ships can be used for flights to large and small planets, to meet with asteroids or comets, and to form special orbits of motion in the vicinity of the Sun or the Earth. New technologies should bring visible results in the creation of space engines based on the direct use of an unlimited source of solar energy. Over the years, numerous variants of motion patterns and new possible forms of solar sails have appeared. The technology for large-scale solar reflector designs is in its infancy.

After constructing and orbiting such mirrors of certain proportions, we obtain a self-adjusting systemic orientation with respect to the Sun for coplanar or spatial trajectories. More sophisticated options and models make it possible to control the orbital and rotational movements of the spacecraft while in motion using a solar sail.

One of the most difficult problems of mankind at the turn of the twentieth/twenty-first centuries was the problem of providing energy. One of the most sensible ways to save energy resources in space is to develop renewable or "perpetual" energy sources. Such promising energy resources primarily include the energy of the sun's rays, both thermal for filling special devices and mechanical for the formation of additional pressure forces on the surface of spacecraft or special installations of "solar sails."

The latest version involves the use of the pressure of the sun's rays on all the bodies encountered in the luminous flux. This "eternal", and at the same time environmentally friendly type of energy resources is also beneficial because it does not require any expenses for its transportation to the place of consumption.

The forms of using the thermal energy of the Sun on Earth are widely known, but the problem of using a very small mechanical pressure of the light flux for the purposes of the so-called "space navigation" turns out to be much more complicated. Flying in space under a solar sail is just a real embodiment of the idea of full or partial replacement of the energy of jet engines with the "donated" energy of the sun's rays, the pressure of which on a mirror reflective sail is able to create, albeit a small, but quite tangible thrust force in space for a spacecraft... With the help of a solar sail, you can determine or change the direction of movement in orbit or perform complex gravity assist maneuvers around large planets.

Opening the sail in orbit will cause the light pressure to partially compensate for the sun's gravity. If the gravitational field of attraction is supplemented by the field of forces of light repulsion, then for problems of celestial mechanics one can speak of a mathematical model of the photogravitational field [3–5]. The acceleration imparted to the spacecraft by the flow of solar rays depends on the ratio of the sail area to the mass of the entire structure. With such modeling, it is sufficient to restrict oneself to taking into account two main forces: the gravitational interaction of bodies F_g and the light pressure F_p on the body from the flow of solar radiation. But for problems of motion in the vicinity of the Earth, it is necessary to take into account the features of the Geopotential and other forces (atmospheric resistance, the influence of other bodies, etc.).

The use of a solar sail will provide the spacecraft with a low-thrust engine, which has an almost unlimited supply of fuel. However, it has a disadvantage: Unlike jet engines, we cannot use its thrust in an arbitrary direction with the same efficiency. The resulting force is determined by the position of the spacecraft in space, as well as by the orientation of all elements of the solar sail relative to the attracting centers and centers of radiation. It is necessary to specifically orient the sail to obtain the desired change in orbital parameters.

Problems of motion control with a solar sail lead to the study of mathematical models of dynamics in photogravitational fields of orbital motion and problems of control over the rotation of the entire spacecraft complex relative to the center of mass [2–21].

The influence of the main forces determines not only the orbital and rotational motion, but can also be used to implement control during interplanetary flights and maneuvering in the sphere of action of the next planet, or to stabilize the spacecraft orientation during its orbital motion. This will make it possible to form the direction and thrust of such a solar-powered engine with an unlimited margin (solar sails with a good mirror surface for reflecting the light flux), which depends on the distance to the source, and the area and shape of the surface of the spacecraft sail elements [3, 9].

3. The main problems of photogravitational celestial mechanics

3.1 Photogravitational celestial mechanics (PhCM) with one radiation source

(simulation of motion in the framework of the two-body problem or the restricted. three-body problem for the solar system)

1. Heliocentric motions: The sun as a point source of radiation. The sun as a non-rotating extended source. The sun as a rotating extended source. Darkening effect toward the edge. Energy luminosity of the Sun (numbers), "solar constant" in the Earth's orbit.

2. Heliocentric photogravitational problem of two bodies (beta-meteoroids)

3. Photogravitational limited three-body problem/evolution of orbits taking into account light pressure in special cases:

   a. Sun-planet-meter asteroid,

   b. Sun-comet-comet tail particle

4. Libration points of the photogravitational three-body problem: position, behavior, and stability of libration points in a plane-bounded problem in the presence of resonances.

5. Nonstationary modifications of photogravitational problems with variable physical parameters in the framework of the dynamics of bodies with variable mass.

6. Heliocentric flights of the solar sail to the Sun, large planets, asteroids, comets, planetary satellites. Interplanetary flights over ellipses of the Hohmann-Tsander type in a photogravitational field.

7. A chain of resonant encounters and repeated hyperbolic gravity assist maneuvers on the Earth-Venus or Venus-Sun flight.

8. The exit of a spacecraft with a solar sail from the Solar System, a gravity assist of a spacecraft with a sail near the rotating Sun, taking into account the relativistic effects (Kerr and Schwarzschild metrics, Lense-Thirring effect). Flight to the focus of the Sun's gravitational lens—550 AU from the Earth.

3.2 Geocentric movements within the photogravitational celestial mechanics

1. Theory of perturbed satellite motion under solar radiation pressure.

2. Controlled low-thrust solar sail flights.

   a. Sailing geocentric separation. Spinning to the Moon inside the Earth's sphere of action.

   b. Environmental aspects of light pressure: a solar sail or a bundle of sails as a solar shield (shield) against global warming, a passive relay, an orbital illuminator (reflector) of near-polar regions, an element of a cable system for changing the orbit of a dangerous asteroid to deflect it from the Earth. Deviation of an asteroid from the orbit of a dangerous approach to the Earth due to an artificial increase in its albedo (black-and-white control coatings), the use of axial spinning of the asteroid due to moments of light pressure forces (the Yarkovsky-Radzievsky effect, i.e., the transverse rotational thermodynamic effect).

   c. Solar sail in a perturbing Coulomb field—the effect of an electric static charge induced on the sail surface on its strength characteristics and on the dynamics of movement of a vehicle with a sail along a Zander trajectory, charging a thin sailing film in space plasma.

3.3 Photogravitational celestial mechanics with two sources of radiation

(movement of interstellar dust within the limited three-body problem)

1. Points of libration: positions, behavior, and stability

2. Instability of circumstellar dust complexes and cloud clusters of microparticles in binary and multiple stellar systems.

3.4 Photogravitational celestial mechanics in the solar system

(objects of research—objects of high windage)

1. Natural small bodies in the solar photogravitational field.

   a. Orbital motions of small bodies taking into account the pressure of solar radiation: dust particles (beta-meteoroids), dust particles of a comet tail, small (meter) asteroids with high albedo, dust complexes around major planets.

   b. Rotational movements under the influence of solar radiation pressure (the thermodynamic effect of Yarkovsky-Radzievsky, thermal inertia of a rotating body).

   c. Orbital relativistic aberration Poynting-Robertson effect: a spiral orbital twist of particles toward the Sun, followed by a catastrophic fallout on it.

   d. Dynamics of emission of dust particles from the head of a comet under the action of light pressure (the parachute effect according to Radzievsky, the movement of particles from the nucleus to the tail in the framework of the limited three-body problem).

2. Artificial celestial bodies with high windage: spherical and ellipsoidal inflatable satellites, spacecraft with solar sails, large transformable space structures, etc. Calculation of light pressure on bodies of various shapes—regular and irregular. Spherical AES—the theorem about the light pressure on an ideal sphere.

   a. Controlled flights of spacecraft with low thrust of the solar sail. Closed interplanetary flights with a sail, near-solar maneuvers.

   b. Near-planetary and circumsolar gravity-assist maneuvers with a solar sail.

   c. Orbit correction (satellite retention in Geo Stationary Orbit).

   d. Rotational movements: 3-axis stabilization, orientation, and control in space under the influence of the moment of radiation pressure forces.

   e. A solar sail as an element of a cable system for transporting a dangerous asteroid from orbit.

4. Mathematical models and equations of motion

For many years, observing the world around us, scientists have tried to understand, describe, or predict the dynamics of various objects: the movement and flight of a stone or spear after a throw, and the movement of stars or comets in the sky. The properties, principles, and patterns of processes were gradually formed.

Aristotle formed the general principles of movement, created a theory of the movement of the celestial spheres, and considered the source of movements to be forces caused by external influences.

Kepler, based on the results of processing tables of observations of planetary motion, taking into account the **Copernican** hypothesis, discovered the laws of planetary motion, proposed convenient parameters for describing possible orbits, and laid the foundation for celestial mechanics.

Laplace, in his multivolume monograph, completed the creation of celestial mechanics based on the law of universal gravitation.

The use of mathematical modeling methods makes it possible to form various equations of dynamics, to study solutions or properties for a selected set of bodies, taking into account the main forces of interaction, when all other conditions are considered insignificant. This allows more complex variants, conditions, and models to be studied later, as well as to obtain new results.

For the forces of attraction, **Galileo's** experiments made it possible to introduce a uniform gravitational field and motion with constant acceleration $g = \text{const}$:

$$mw = F = mg \tag{1}$$

The law of universal gravitation, when describing the motion of bodies in absolute space, allowed **Newton** to obtain mathematical models of the gravitational field and exact solutions that confirm **Kepler's** laws for the motion of the planets of the solar system.

$$m_1 w_1 = F_1 = -f \frac{m_1 m_2}{r^3} r_{12} \tag{2}$$

$$m_2 w_2 = -f \frac{m_1 m_2}{r^3} r_{21} = F_2 = -F_1 \tag{3}$$

Considering a system of only two interacting bodies, we can write down the equations of motion, which, taking into account Newton's laws, determine the properties for the center of mass of the system.

$$m_1 w_1 + m_2 w_2 = 0, \quad m_1 v_1 + m_2 v_2 = \text{const} \tag{4}$$

Galileo's principle of relativity asserts the existence of a frame of reference in which the center of mass retains its position, and two bodies move in their orbits around it. The motion in the two-body problem determines the gravitational parameter μ of the central body. Equations for any selected system of bodies can be written in a similar way:

$$w_i = -\frac{\mu}{r^3} r_i, \quad \mu = f(m_0 + m_i). \tag{5}$$

The motion of a body in the Cartesian coordinate system relative to the central body can be written by equations

$$\frac{d^2 x_i}{dt^2} = -\frac{\mu}{r^3} x_i, \quad i = 1,2,3 \tag{6}$$

The parameters that determine the action of the central force are formed on the basis of indirect observations for the coefficient of the gravitational constant and the masses of the bodies involved. And if you need to write down equations for a planetary problem, where formally there are ten planets and hundreds or thousands of small bodies of the solar system, then with what accuracy are all parameters represented?

The principle of determinism makes it possible to obtain coordinates at any moment in time for such a model with constant parameters and orbital elements.

The motion in the central gravitational field has a solution, which in the absence of disturbing forces is determined by the initial values of the radius vector, the velocity vector, and the gravitational parameter of the central body. This determines the constant Keplerian elements $\mathbf{k} = (a, e, i, \Omega, \omega, M_0)$, which allow calculating the Cartesian coordinates $x_i(t)$ and velocities $v_i(t)$ for unperturbed motion at an arbitrary moment in time t by formulas [19]:

$$\left\{\begin{array}{l} x_1 = r(\cos u \cos\Omega - \sin u \sin\Omega\cos i) \\ x_2 = r(\cos u \sin\Omega + \sin u \cos\Omega\cos i), \\ x_3 = r\sin u \sin i, \end{array}\right. \tag{7}$$

$$\left\{\begin{array}{l} v_1 = \alpha(\cos u \cos\Omega - \sin u \sin\Omega\cos i) \\ -\beta(\sin u \cos\Omega + \cos u \sin\Omega\cos i), \\ v_2 = \alpha(\cos u \sin\Omega + \sin u \cos\Omega\cos i) \\ -\beta(\sin u \sin\Omega + \cos u \cos\Omega\cos i) \\ v_3 = \alpha\sin u \sin i + \beta\cos u \sin i. \end{array}\right. \tag{8}$$

Here

$$r = a(1 - e\cos E), \quad p = a(1 - e^2),$$

$$\alpha = \sqrt{\frac{\mu}{p}}er^{-1}\sin\vartheta, \quad \beta = \sqrt{\mu p}r^{-1}, \tag{9}$$

The time of movement between two points of the orbit can be determined from the equation, which is called the Kepler equation

$$E - e\sin E = M_0 + n(t - t_0) = M. \tag{10}$$

Osculating elements $\mathbf{k}(t) = (a, e, i, \Omega, \omega, M_0)$ can be used to describe the motion with allowance for perturbations, when the spacecraft orbital elements are functions of time. Differential equations can be used, where the right-hand sides are determined by the current values of the elements and the projections of the disturbing accelerations on the axis of the orbital coordinate system.

$$\left\{\begin{array}{l} \dfrac{da}{dt} = 2a^2\left(e\sin\vartheta P_1 + pr^{-1}P_2\right), \\[2mm] \dfrac{de}{dt} = p(\sin\vartheta P_1 + \cos\vartheta P_2 + \cos E P_2), \\[2mm] \dfrac{di}{dt} = r\cos\vartheta P_3, \quad \dfrac{d\Omega}{dt} = r\sin u \sin^{-1}i \, P_3, \\[2mm] \dfrac{d\omega}{dt} = e^{-1}[(r+p)\sin\vartheta P_2 - p\cos\vartheta P_1] - \cos i\dfrac{d\Omega}{dt}, \\[2mm] \dfrac{dM_0}{dt} = \sqrt{e^{-2}-1}[(p\cos\vartheta - 2er)P_1 - (r+p)\sin\vartheta P_2]. \end{array}\right. \tag{11}$$

In the general case, the vector projections on the radial and transverse directions will affect the change in the parameters of the motion orbit. Projection to the normal to the orbital plane will allow you to change its inclination relative to its original position.

The third and fourth equations of system (6) show that the initial position of the orbit is preserved in the absence of a projection of the disturbing forces on the normal to the plane.

If the mathematical model defines the equations, and the properties of the solutions suit the researchers, then great. Analytical or numerical methods make it possible to refine objects and predict movement.

The discovery and experimental confirmation of the effect of light pressure by **Lebedev** makes it possible to use the photogravitational field, which introduces a correction (reducing the force of attraction to the central body by the amount of light pressure on the surface of the body) $\mu_f = \mu_g - \mu_p = \mu_g(1 - \delta_p)$. For asteroids with similar parameters, the coefficients of light pressure will be close.

A new level of influence of light pressure appeared after Tsander 's idea of spacecraft flights under a solar sail, where the parameters of the ratio of body surface area and mass are significantly different. Additionally, it becomes possible to change the direction of the thrust vector, which takes into account the position of the sail and the properties of the mirror surface that reflects the streams of sunlight. In this case, the central gravitational field is supplemented by a special governing force. In the projection on the axis of the Cartesian coordinate system, you can write (2).

The equations can be written, as suggested by **Euler**, using **Kepler's** osculating elements to use the projections of forces on the axis of the orbital coordinate system to change the parameters of the orbit of motion over time as a feedback control.

The problems of motion of the CASP in the vicinity of the Earth or another planet can be investigated on the basis of the limited problem of three bodies, supplemented by a special control force, depending on the relative position of the two main bodies: the Sun and the planet.

The ability to control the orientation of the CASP and individual elements of the sails system has a special character to form the best control of the main thrust vector and the moment of forces for turning the CASP relative to the center of mass or maintaining the desired position.

The main problem of the relative motion of two bodies under the action of gravitational forces of interaction (without taking into account other perturbing forces) is reduced to the equations of the central force field, describing the movement of a material point along an elliptical trajectory, in the focus of which is the attracting center of the main body. The magnitude of the force F_g depends on the square of the distance and the gravitational parameter. In the projection on the axis of the Cartesian coordinate system, you can write Eqs. (6).

If we restrict ourselves to motion while maintaining the initial plane of the orbit, then we can use polar coordinates $r(t), \phi(t)$ and projections of forces on the radial and transverse directions

$$\frac{d^2 r}{dt^2} + \frac{\mu}{r^2} - r(\dot{\varphi})^2 = P_1, \quad \frac{d}{dt}\left(r^2 \dot{\varphi}\right) = P_2, \tag{12}$$

The motion in the central gravitational field has a solution, which in the absence of disturbing forces is determined by the initial values of the radius vector, the velocity vector and the gravitational parameter of the central body.

The discovery and experimental confirmation of the effect of light pressure by **Lebedev** makes it possible to use the photogravitational field, which introduces a correction (reducing the force of attraction to the central body by the amount of light pressure on the surface of the body)

$$\mu_f = \mu_g - \mu_p = \mu_g(1 - \delta_p).$$

For asteroids with similar parameters, the coefficients of light pressure will be close.

A new level of influence of light pressure appeared after Tsander's idea of spacecraft flights under a solar sail, where the parameters of the ratio of body surface area and mass are significantly different. Additionally, it becomes possible to change the direction of the thrust vector, which takes into account the position of the sail and the properties of the mirror surface that reflects the streams of sunlight. In this case, the central gravitational field is supplemented by a special governing force.

In the projection on the axis of the Cartesian coordinate system, you can write (3).

The equations can be written, as suggested by Euler, using Kepler's osculating elements to use the projections of forces on the axis of the orbital coordinate system to change the parameters of the orbit of motion over time as a feedback control.

The problems of motion of the CASP in the vicinity of the Earth or another planet can be investigated on the basis of the limited problem of three bodies, supplemented by a special control force, depending on the relative position of the two main bodies: the Sun and the planet.

The photogravitational field in the three-body problem for motion in the vicinity of the Earth can be considered a combination of the central field or geopotential and the disturbing action of the Sun's gravitational force and the forces of light pressure. A modification of the restricted three-body problem is obtained, taking into account the additions that can form control forces for the implementation of the CASP flight program along a given trajectory in the vicinity of the Earth or for spinning the initial orbit in the case of the problem of getting out of the sphere of the Earth's gravity for flights into the distant expanses of the Solar system.

If motion in the vicinity of the Earth is considered, then the directions of the two main forces do not coincide, but it can be assumed in a first approximation that the luminous flux determines an almost constant pressure force, collinear with a straight line that passes through the two main bodies of the system. The position of the sail plane allows you to form the direction of the control force $\mathbf{P}(t,u)$ to change the trajectory or stabilize in the vicinity of the singular libration points. Then, you can use the equations of motion in the framework of the limited circular problem of three bodies

$$\ddot{x} - 2\eta \, \dot{y} = \frac{\partial U}{\partial x} + P_1, \quad \ddot{y} + 2\eta \, \dot{x} = \frac{\partial U}{\partial y} + P_2, \quad \ddot{z} = \frac{\partial U}{\partial z} + P_3. \tag{13}$$

The position of the center of a spacecraft of infinitely small mass relative to the main bodies (Earth and Sun) of mass $\mu < < 1$ and $(1 - \mu)$ in a rotating barycentric Cartesian coordinate system is determined by the radius vectors

$$\vec{r} = (x, y, z), \quad \vec{r}_1 = (x + \mu, \ y, \ z), \quad \vec{r}_2 = (x - 1 + \mu, \ y, \ z). \tag{14}$$

The force function of the gravitational interaction has the form

$$U = \frac{1}{2}\eta^2 (x^2 + y^2) + \kappa^2 \left(\frac{1 - \mu}{r_1} + \frac{\mu}{r_2} \right). \tag{15}$$

Here, η is the constant angular velocity of rotation of the coordinate system relative to the center of mass of the system together with the main bodies. Thus, there is an additional simplification at $P_1 = const$, $P_2 = P_3 = 0$. When moving in the vicinity of the Earth, for an approximate solution, one can use the intermediate orbits of the Hill problem [17], including when studying the stability of motion in the vicinity of libration points.

5. Control and stabilization capabilities

The ability to control the orientation of the CASP and individual elements of the sails system has a special character to form the best control of the main thrust vector and the moment of forces for turning the CASP relative to the center of mass or maintaining the desired position.

Heliocentric motion of a spacecraft with a solar sail during flights to the Sun, planets, asteroids, or comets; and also to create special orbits of motion in the vicinity of the Sun, taking into account the force of light pressure, leads to the equations of motion in the form of a system

$$\frac{d^2 x_i}{dt^2} = -\frac{\mu}{r^3} x_i + \frac{\partial U}{\partial x_i} + f_i(x) + u_i(t, x), \quad i = 1,2,3, \tag{16}$$

where the notation is used: x_i are the Cartesian coordinates of the spacecraft, r is the modulus of the radius vector, μ is the gravitational parameter of the central body, the function U is determined by the influence of disturbances from potential forces, and the functions $f_i(x)$, $u_i(t, x)$ are the components of the acceleration vectors non-potential forces and the vector control contribution $u(t, x)$, including the action of light pressure forces or jet engines on active sections of motion, when expanded on the axes of the orbital coordinate system [20]. In this case, the forces of light pressure on the elements of the sail system and the moments relative to the center of mass of the system determine the vector quantities:

$$\mathbf{F} = \sum \mathbf{F}_i = \sum k_i S_i \frac{b\ (\theta_i)}{r^2} \mathbf{n}_i\ (\theta_i), \quad \mathbf{M} = \sum_i \rho_i \times \mathbf{F}_i(\theta_i). \tag{17}$$

The contribution of the light pressure is determined by the angle of deviation of the normal vector n from the direction of the flux of the sun's rays e. If the flat mirror sail is angled θ_i to the beams, then the transmitted pulse will be directed almost perpendicular to the reflective surface. The photons will retain a part of the impulse directed parallel to the sail, so that the sail will get less than with full opening to the rays. The magnitude of the light pressure decreases, and the direction will almost coincide with the normal to the sail, laid down from its shadow side. By turning the sail, we get the opportunity to change the direction of the thrust vector and control the spacecraft. However, this changes the value. If the normal of a flat sail is perpendicular to the stream of rays, then the sail exerts no thrust at all. In the general case, the vector projections onto the radial and transverse directions will affect the change in the parameters of the motion orbit. Projection to the normal to the orbital plane will allow you to change its inclination relative to its original position. Acceleration also depends on the ratio of the sail area S to the mass of the entire structure and on the surface properties or surface reflection coefficient. Here, the designations are additionally used when summing over all structural elements.

The design features of the spacecraft in the problems under consideration make it possible to use a solar sail to control motion. By turning the sail, taking into account the pressure of sunlight, it is possible to control the orbital dynamics and the relative position of the spacecraft in space, as well as perform stabilization.

For rotational motion relative to the center of mass, the kinetic energy of the body is determined by the moments of inertia and angular velocity

$$T = \tfrac{1}{2}\left(I_x \omega_x^2 + I_y \omega_y^2 + I_z \omega_z^2\right).$$

Differential equations can be written under the action of the moment of forces in the form

$$
\begin{cases}
I_x \dot{\omega}_x = (I_y - I_z)\omega_y\omega_z + M_x, \\
I_y \dot{\omega}_y = (I_z - I_x)\omega_x\omega_z + M_y, \\
I_z \dot{\omega}_z = (I_x - I_y)\omega_y\omega_x + M_z
\end{cases}
\tag{18}
$$

Taking into account the effect of light pressure on the spacecraft sail leads to the appearance of special stability conditions that can be used to control the movement. Correctly chosen shape of the sail allows you to keep the spacecraft in the desired position. The effect of disturbances can be compensated for by changing the size or reflective properties of the elements of the sail of the spacecraft, as well as their relative position. This creates the additional moments of force that can be used as controls.

If we introduce into consideration the control moments u_i ($i = 1,2,3$), relative to the main axes of inertia, and use the projections of the angular momentum as unknowns $x_7 = I_x\omega_x$, $x_8 = I_y\omega_y$, $x_9 = I_z\omega_z$, then the system of equations of motion (1) and (4) for the considered set of generalized coordinates can be represented in the normal form:

$$
\begin{cases}
\dot{x}_i = x_{i+3}, \quad i = 1, 2, 3, \\
\dot{x}_{i+3} = -\mu \ r^{-3} \ x_i + f_i(t, \ x_i, \ u_i), \\
\dot{x}_7 = \beta_1 \ x_8 x_9 + u_4, \\
\dot{x}_8 = \beta_2 \ x_9 x_7 + u_5, \\
\dot{x}_9 = \beta_3 \ x_8 x_7 + u_6, \\
\beta_1 + \beta_2 + \beta_3 = 0.
\end{cases}
\tag{19}
$$

In the case of possible oscillations [13], while maintaining the orientation of one of the main axes orthogonal to the plane of motion with angular orbital velocity, the change in the angular momentum taking into account the action of the geopotential can be investigated using the equation

$$
I_z \dot{\omega}_z = -k\omega_0^2 (I_x - I_y) \sin \phi = M_z.
\tag{20}
$$

Taking into account the additional force leads to new, different from the classical, formulations of optimal control problems, and to new mathematical models. In these models, additional equations appear and other control functions are considered. To solve such problems, the methods of Pontryagin or the Bellman equation are used. There are analytical and numerical methods of research and analysis of the main properties of new equations, which allow obtaining exact or approximate solutions that deliver an extremum to the quality criterion and satisfy the necessary conditions [8, 21–23].

6. Inverse problem of photogravitational dynamics

Considering the inverse problem for finding forces for a known or given motion in the case of two interacting bodies, we find the control accelerations that can realize the trajectory in the central field

$$
f_i = \frac{d^2 x_i}{dt^2} + \frac{\mu}{r^3} x_i = u_i(t), \quad i = 1,2,3,
\tag{21}
$$

when the parametric functions $x_i(t)$ are predefined for $t \in [0,T]$.

The results of solving the inverse problem of dynamics are presented as examples. Simulation of closed trajectories $x = r_0 \cos^2 t$, $y = r_0 \sin t \cos t$, $z = h_3 \sin t$ of a spacecraft with a controlled solar sail is presented to reach the heliopolar regions (**Figure 1**, Viviani curve), and fly over the North and South poles of the Sun and return to near-earth orbit [2]. Assuming that the required control accelerations are created by the position of the solar sail [5], we write them in the form

$$\begin{cases} u_1 = -\dfrac{q}{r^2} \cos \gamma_1 \cos \gamma_2, \\[2mm] u_1 = \dfrac{q}{r^2} \sin \gamma_1 \cos \gamma_2, \\[2mm] u_1 = \dfrac{q}{r^2} \sin \gamma_2 \end{cases} \tag{22}$$

where γ_1 and γ_2 are the angles of the three-dimensional orientation of the normal vector to the shadow side of the mirror sail on both sides, and the dimensionless factor k depends on the solar radiation pressure on the surface of the spacecraft sail (**Figure 2**). From here, we obtain the formulas for finding the angles and the construction of the control.

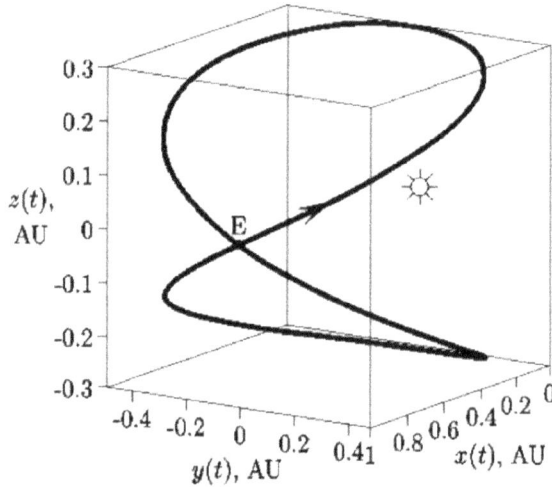

Figure 1.
The Viviani curve is slightly flattened in the vertical direction at $h_3 = 0.3 r_0$.

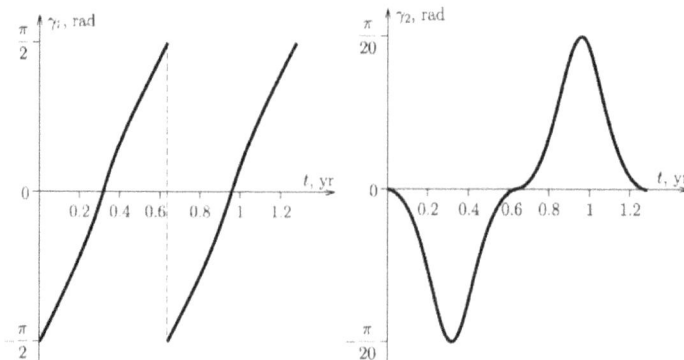

Figure 2.
Graph of γ_1 and γ_2 for Viviani curve at $h_3 = 0.3 r_0$, $T_3 = 1.03$ years.

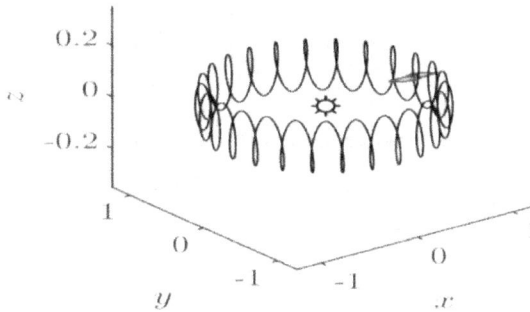

Figure 3.
The spiral trajectory is defined by expressions in the orbital coordinate system.

The spiral trajectory (**Figure 3**) of the possible motion of the CASP is determined by the equations

$$\begin{cases} x = r_0 \cos \omega_0 t, \\ y = r_0 \sin \omega_0 t + r_1 \cos \omega_1 t, \\ z = r_1 \sin \omega_1 t. \end{cases} \tag{23}$$

The helical trajectory can be supplemented by the selected law of control over the position of the sail elements.

The solution of the inverse problem of dynamics when moving along a given trajectory makes it possible to obtain an initial approximation for control and to evaluate the possibility of implementing a system of spacecraft sail elements for the selected model. With this solution, you can also refine the rotation or steering of the entire structure with respect to the center of mass.

7. Conclusion

The functional features of the motion of a spacecraft with solar sails are considered. The main tasks of photogravitational celestial mechanics are listed, and the possibilities of their mathematical modeling are given. The main differences in this area of research are sophisticated spacecraft models, supplemented by solar sails of various shapes, sizes, and physical properties. Taking into account all the characteristics of the sail complicates the process of mathematical modeling of the spacecraft, but it is possible with the help of classical analytical dynamic tools.

The dynamic aspect of the simulation is to take into account the solar pressure force on the sail. The presence of this force determines the specifics of the formulation of control tasks for such spacecraft, closely related to their design features. Despite the differences between such formulations from the usual optimal control problems, their solution can be obtained by classical optimization methods.

Author details

Elena Polyakhova and Vladimir Korolev*
Saint Petersburg State University, Saint Petersburg, Russia

*Address all correspondence to: v.korolev@spbu.ru

IntechOpen

References

[1] Tsander FA. The problem of flight with the help of jet vehicles. Collection of articles edited by M. K. Tikhonravov. Moscow: Oborongiz; 1947. p. 237 (in Russian)

[2] Polyakhova EN. Space flight with a solar sail: Problems and prospects. Moscow: Nauka, 1986; 2nd edition, Moscow: Book House "LIBROCOM", 2011, 3rd ed. ster.: Book house. Librocom (URSS). 2018. (in Russian)

[3] Polyakhova EN, Shmyrov AS. Physical model of the forces of light radiation pressure on the plane and sphere. In: Vestnik SPbSU. Ser. 1.vol. 2. St. Petersburg: St: Petersburg State University Publishing House; 1994. pp. 87-104. (in Russian)

[4] Polyakhova E N. To the centenary of photogravitational celestial mechanics. In: Vestnik SPbSU. Ser. 1. Issue 4. St. Petersburg: St. Petersburg State University Publishing House; 2004. pp. 89-118. (in Russian)

[5] Polyakhova EN, Starkov VN, Stepenko NA. Flights of a spacecraft with a solar sail outside the ecliptic plane. In: Stability and Control Processes. Materials of the Conference. Petersburg: St. Petersburg State University Publishing House; 2015. pp. 91-92 (in Russian)

[6] Polyakhova EN, Grinevitskaya LK, Construction of an Approximate Solution to the Equations of Geocentric Motion of a Spacecraft with a Solar Sail. Vol. 7. Leningrad: Book House LGU, Vestnik Leningrad University; 1973. pp. 134-143 (in Russian)

[7] Perezhogin AA, Tureshbaev AT. On the stability of triangular libration points in the photogravitational problem of three bodies. Astronomical journal. 1989;**66**:859-865

[8] Korolev VS, Polyakhova EN. Pototskaya IY, Stepenko NA. Mathematical models of a solar sail spacecraft controlled motion. In: Proceedings of 2020 15th International Conference on Stability and Oscillations of Nonlinear Control Systems (Pyatnitskiy's Conference). vol. 2020. STAB. 2020. p. 9140472

[9] Kirpichnikov SN, Kirpichnikova ES, Polyakhova EN, Shmyrov AS. Planar heliocentric roto-translatory motion of a spacecraft with a solar sail of complex shape. In: Celestial Mechanics and Dynamical Astronomy. Vol. 63(3–4). 1995. pp. 255–269

[10] Koblik VV, Polyakhova EN, Sokolov LL, Shmyrov AS. Controlled solar sailing transfer flights into near-sun orbits under restrictions on sail temperature. Cosmic Research. 1996; **34**(6):572-578

[11] Polyakhova E, Pototskaya I, Korolev V. Problems of control motion and stability of solar sail spacecraft. In: 2017, International Multidisciplinary Scientific Geo Conference Surveying Geology and Mining Ecology Management, SGEM. Vol. 17 2017. pp. 939-946. 62 Pages

[12] Polyakhova E, Pototskaya I, Korolev V. Spacecrafts control capabilities via the solar sail in the photogravitational fields. In: 2019, 19th International Multidisciplinary Scientific Geoconference, SGEM. Vol. 19(6.2). 2019. pp. 699–706.

[13] Polyakhova E, Korolev V. The solar sail: Current state of the problem. In: 2018, 8th Polyakhov's Reading: Proceedings of the International Scientific Conference on Mechanics. Vol. 1959. American Institute of Physics. p. 040014

[14] Radzievsky VV. Bounded problem of three bodies taking into account light

pressure. In: Astronomical Journal. 1950;**30**(4):249–256

[15] Rauschenbach BV, Tokar EN. In: Nauka M, editor. Control of Spacecraft Orientation. 1974 (in Russian)

[16] Tureshbaev AT. Nonlinear stability analysis of triangular libration points in the generalized photogravitational plane problem of three. Mathematical Journal, Almaty. 2010;**10**. (37):101-106. No. 3

[17] Subbotin MF. Introduction to Theoretical Astronomy. Moscow: Nauka; 1968. p. 800 (in Russian)

[18] Polyakhova E, Starkov V, Stepenko N. Flights of a spacecraft with a solar sail out of ecliptic plane. In: AIP Conference Proceedings 8th Polyakhov's Reading. 2018. p. 040017

[19] Beletsky VV. Motion of an Artificial Satellite about its Center of Mass. Moscow: Nauka, (in Russian); 1965

[20] Polyakhova E, Korolev V, Problem of spacecraft control by solar sail. In: Proceedings of 2016 International Conference "Stability and Oscillations of Nonlinear Control Systems" (Pyatnitskiy's Conference), STAB. Institute of Electrical and Electronics Engineers Inc; 2016. p. 7541214

[21] Novoselov VS, Korolev VS. Analytical Mechanics of a Controlled System. St. Petersburg: St. Petersburg State University Publishing House; 2005 p. 298

[22] Tikhonov AA, Aleksandrov AY. Electrodynamic stabilization of earth-orbiting satellites in equatorial orbits. Cosmic Research. 2012;**50**:313-318

[23] Fedor NNI. Aleksandrovich Bredikhin. Moscow: Leningrad, "Science"; 1964. p. 254

Section 2

Application of the Gravity Method in Exploration

Chapter 5

Gravity Anomaly and Basement Estimation Using Spectral Analysis

Mukaila Abdullahi

Abstract

Gravity survey and interpretations play a very vital role more especially in petroleum prospecting. Spectral analysis of gravity anomaly has been successful in the estimation of sedimentary basement. Spectral analysis technique can be used in designing filter for the residual and region separation of complete Bouguer anomaly. The residual gravity anomaly which is of prime importance for applied geophysicists interested in the subsurface features is considered most useful for the interpretation of sedimentary basin. In this chapter, interpretation of the complete Bouguer gravity anomaly, the importance of the separation of the Bouguer gravity anomaly into its residual and regional component is presented. The residual component is considered for the application of the spectral analysis approach.

Keywords: gravity anomaly, regional/residual anomaly, spectral analysis, sedimentary basement, basement estimation

1. Introduction

Geophysics is a field of study that deals with the study of the physical properties of the earth's interior generally by direct measurement on or above the earth's surface in order to find the quantitative value of the earth attraction force on a given buried material below the subsurface on the basis of Newton's law of gravitation. Geology and geophysics appeared difficult for one to define the broader line between the two. But in a broader sense, it can be put that, geology deals with study of the physical properties of the earth by direct observations and analysis of handpicked samples from field (i.e., on the ground). While geophysics involves the study of those buried physical properties of the earth using an appropriate measuring instruments on or above the earth's surface. It also involves the interpretation of such measured data to make inferences about the basement structures. As a matter of fact, the two branches of geoscience are interwoven. For instance, well logs are done for geological interpretation whereas borehole geophysics has to do with such measurements.

In a nutshell, geophysics provides the measuring instruments for the measurement of the composition and structures of the earth's interior. All that comes from underneath the earth's surface to limited certain depths to which boreholes or mine shaft penetrated come from geophysical measurements. The knowledge of the existence and properties of the earth's crust, mantle and the core came from observations of the seismic waves by earthquakes, gravity and magnetic measurements and thermal properties of the earth.

There are practically two related aspects of all geophysical deliberations, "pure" and "applied". First of which deals with the understanding of the dynamics of the earth whereas the second one deals with the economic applications which is of prime importance to mankind. However, for pragmatic design and execution of geophysical survey/exploitation, a perceptive understanding of the structure, evolution of the crust and the uppermost mantle and various processes operating on the earth are essential.

Gravity surveys play vital role in recognition of geological structures such as sedimentary basins, faults, caves and other archeological structures [1–5]. There are varieties of techniques and methods for the interpretation of gravity anomalies over or due to sedimentary basins [6–12]. The structure of the sedimentary basins is often derived from the gravity anomalies with constant density contrast throughout the section of the basins [13]. However, the density contrast of sedimentary rocks is not practically constant [9]. The gravity method for hydrocarbon exploration is firstly used in 1924 in the Gulf coast of the United States and Mexico [14]. Till date, the structures in which hydrocarbons are entrapped exhibit such large density variation when compared to the densities of the surrounding rock formations [15]. Gravity method is very useful in deciding an appropriate location for drilling. For suitable geology of an area, gravity data can provide whether the sedimentary thickness beneath the subsurface is sufficiently thick enough to justify further geophysical investigation. This can be done very easily, since the densities of different sedimentary rock formations are usually lower than those of the basement rocks. Whenever this large contrast exists, it is easier to map out and determine the depth distributions of the sedimentary basins [16].

Gravity data/method is also useful in determining the positions/locations and sizes of key source features/structures in which hydrogeological aquifers, enormous base metals, iron ores, salt domes are entrapped or hosted [17, 18]. Gravity anomaly at long wavelength usually suggests undulations possibly in the topographic interface and the lateral variations in its physical properties (densities). While, short wavelength anomaly may suggest density variations related to the nature of the basin fill. These could encompass the compaction, facies changes and basic to intermediate intrusive [17]. Gravity data can be used to study the internal tectonic and stratigraphic framework, basement and crustal structure. Thus, understanding the structural basement framework, thickness and the physical properties of crust and mantle down to lithosphere is very important more especially in hydrocarbon prospect. Moreover, gravity method is still widely used as an exploration tool to map subsurface geology and estimate ore reserves for some massive ore bodies.

In mineral exploration, gravity method plays more applicability especially in search for ores like Chromite bodies [18]. The density contrast between the chromite bodies and the material that surround them can be so large that they can easily be located through direct gravity measurement. In the same sense, buried channels beneath the earth's subsurface containing gold or uranium can be located since the channel fill is usually less dense than the rock in which it is hosted. Regional gravity studies are also very important in delineating major geological structures like faults/ lineaments in which minerals are probably accumulated [19].

2. Gravity data and regional/residual separation

The complete Bouguer airborne gravity data of part of lower Benue trough in Nigeria (**Figure 1**) as acquired and corrected by Fugro Airborne Surveys (FAS) in 2010. The data is issued by the Nigerian Geological Survey Agency (NGSA), Abuja office, Nigeria. The survey was at a terrain clearance of 80 m, along NE–SW oriented flight lines with 4000 m flight line spacing and gridded at 500 m grid spacing. The complete Bouguer anomaly map (**Figure 1**) represents geological

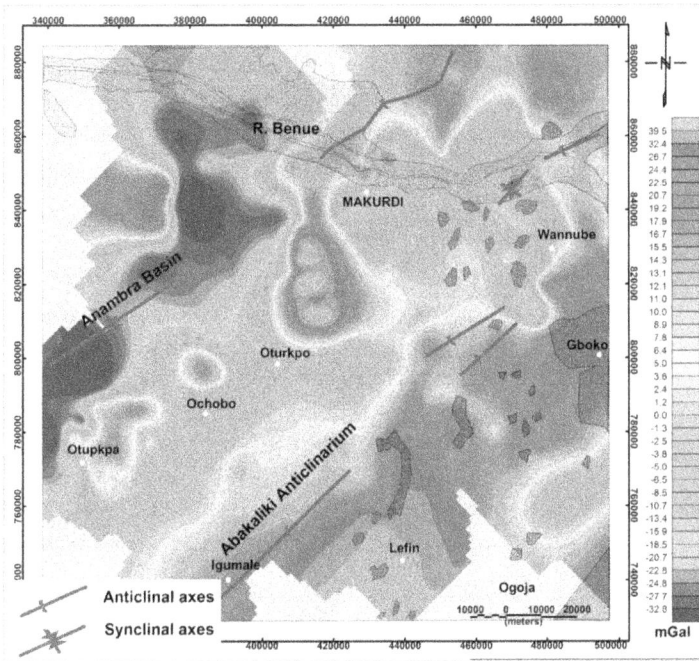

Figure 1.
Complete Bouguer gravity anomaly map. Gboko as well as the basement outcrops east of Ogoja are shown. Intrusions and locations of the anticline and syncline are superimposed.

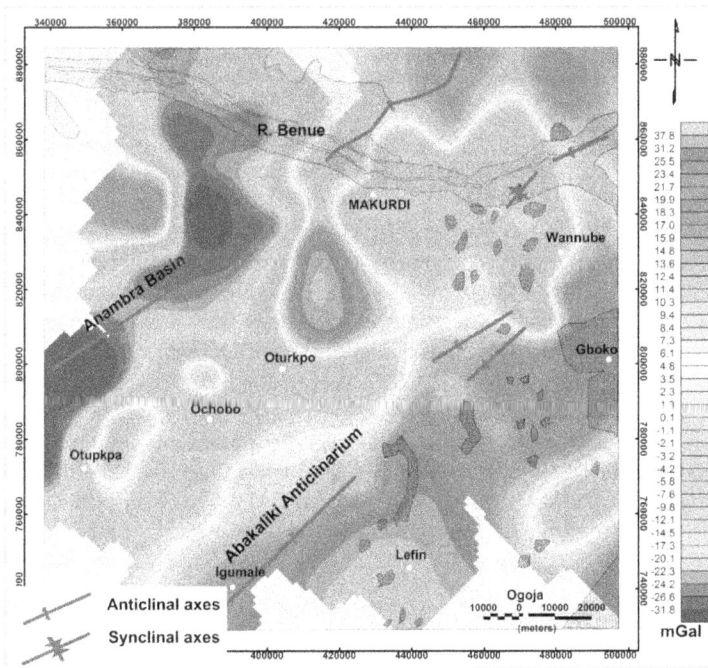

Figure 2.
Regional component of the complete Bouguer gravity anomaly map. Gboko as well as the basement outcrops east of Ogoja are shown. Intrusions and locations of the anticline and syncline are superimposed.

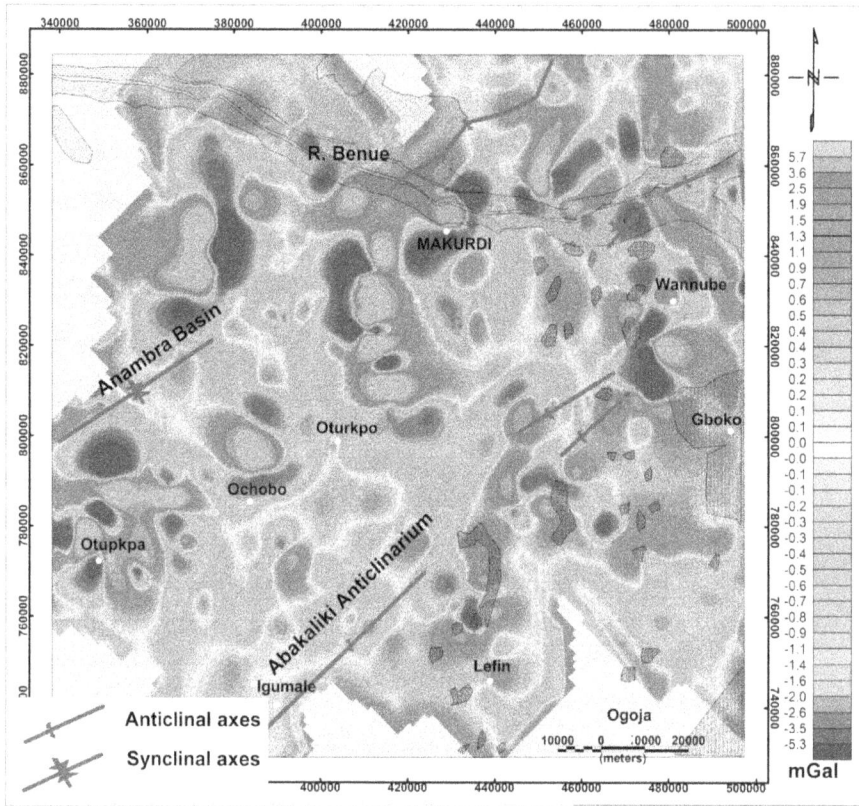

Figure 3.
Residual component of the complete Bouguer gravity anomaly map. Gboko as well as the basement outcrops east of Ogoja are shown. Intrusions and locations of the anticline and syncline are superimposed.

information from both deeper crust and shallower sedimentary thicknesses. In order to outline between the two, regional/residual separation is paramount. The regional anomaly (**Figure 2**) with gravity values between less than −31.8 mGal and 37.8 mGal could be interpreted as gravity information due geological formations in the lower crust possibly to the mantle or even lithosphere. The anomaly map (**Figure 2**) is the result of the Gaussian regional filter (low pass) at 50 km cut-off wavelength in MAGMAP tool of Oasis montaj. The other component of the anomaly (residual anomaly) map (**Figure 3**) which is of prime importance to applied geophysicists is obtained by subtracting the regional gravity anomaly (**Figure 3**) from the complete Bouguer anomaly (**Figure 1**) using Isostatic tool menu. This anomaly map (**Figure 3**) shows shorter wavelengths anomalies that can account for varying depth sedimentary basins (lower gravity values) and probably volcanic structures within the sediments (higher gravity values).

3. Spectral analysis of gravity anomaly

Spectral analysis is a much known technique for the estimation of basement depth to anomalous sources in frequency domain [20–22]. In frequency domain, the depth to the various anomalous sources can be estimated based on their frequency content. Naturally, high-frequency anomalies are due to shallow anomalous sources whereas the low-frequency anomalies are due to anomalous sources that are at greater depths. The average power spectrum of potential field data usually shows a decaying curve with increasing frequency [21]. Spector and Grant [21] method has become an important technique for top depth estimation of anomalous sources in frequency/wavenumber domain [23]. Their concept is based on an ensemble of prisms of frequency-independent randomly and uncorrelated distribution of sources equivalent to white noise distribution. According to Spector and Grant [21] method, the logarithm of the power spectrum generated from horizontal distribution of sources is always directly proportional to $(-2dw)$, where d is the depth to top of sources and w is the radial wavenumber. The depth of the anomalous sources can therefore be estimated directly from the slope of the logarithmic plot of the power spectrum against the wavenumber. The power spectrum of sources could suggest the presence of various horizontal distributions of anomalous sources in the crust pending on the number of segments defined from the power spectrum. The power spectrum is usually not straight, due to the randomly and uncorrelated distribution of anomalous sources assumed in the Spector and Grant [21] method. The random and uncorrelated distribution was assumed due to lack or little knowledge about the depth distribution of the sources. The statistical ensemble of anomalous sources is therefore determined using the following equation

$$\ln[P(w)] = \ln[C] - 2dw = \ln[C] - 4\pi df \qquad (1)$$

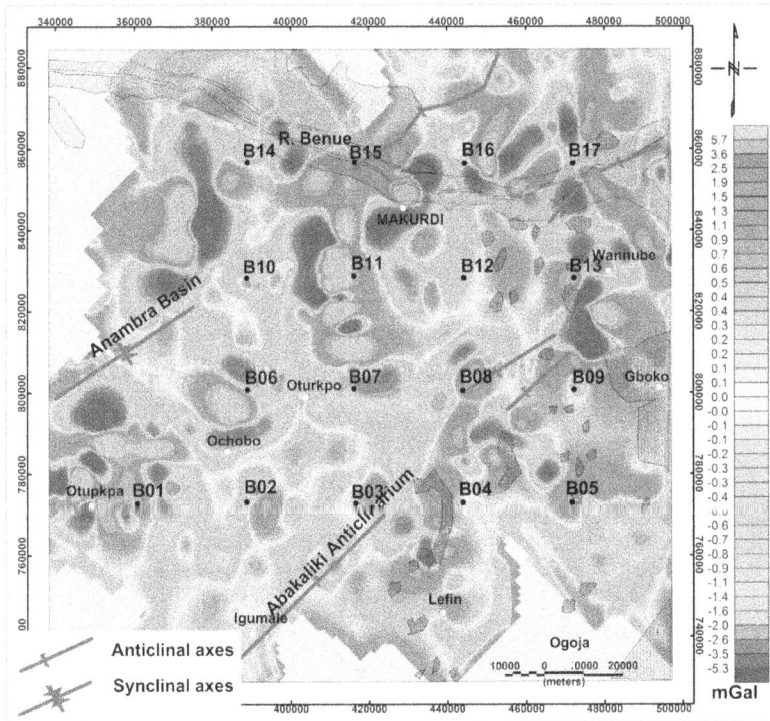

Figure 4.
Residual component of the complete Bouguer gravity anomaly map. Central block (B01, B02, ..., B17) of 55 km x 55 km with 50% overlap are shown for the sedimentary basement estimation. Gboko as well as the basement outcrops east of Ogoja are shown. Intrusions and locations of the anticline and syncline are superimposed.

This equation is analogous to equation of straight line as follows

$$y = mx + C \qquad (2)$$

Comparing the two equations, it can be shown that, the slope (m) power spectrum is given as

$$m = -2d = -4\pi d \qquad (3)$$

$$\text{or depth}, d = -\frac{m}{2} = -\frac{m}{4\pi} \qquad (4)$$

Figure 4 shown, is the residual gravity anomaly map showing the central 55 km x 55 km blocks used for the estimation of sedimentary basement. Seventeen (17)

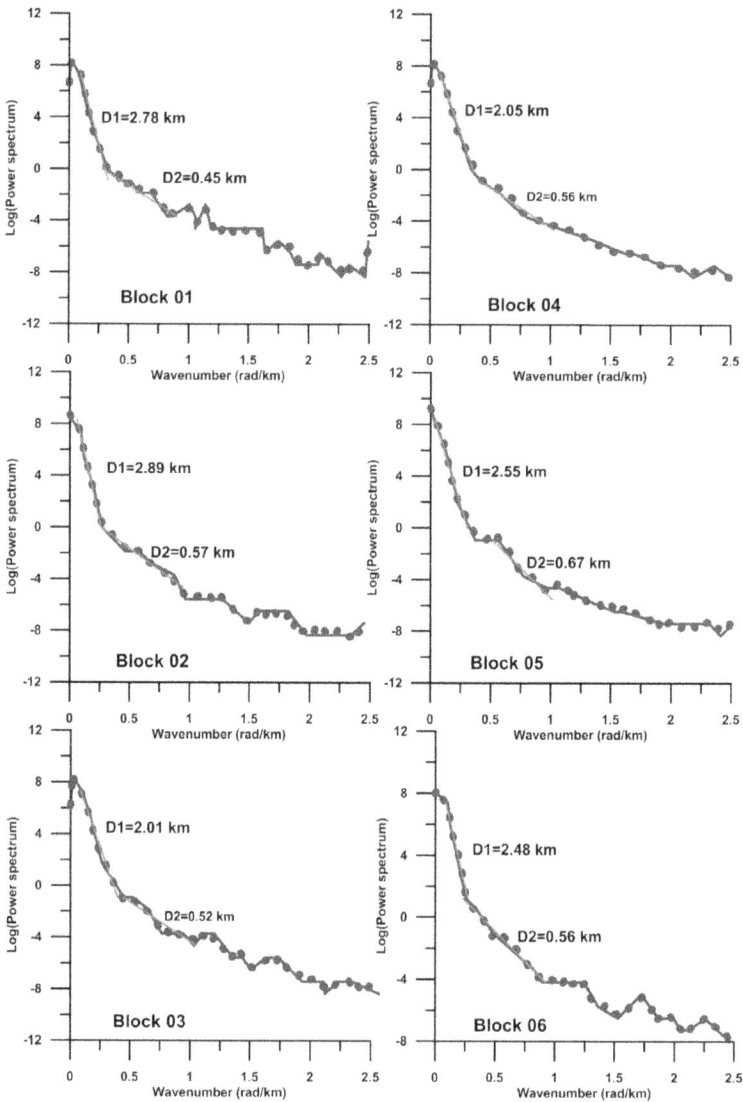

Figure 5.
Power spectra of blocks (B01, B02, ..., B06), showing the depths (D1 & D2) estimated.

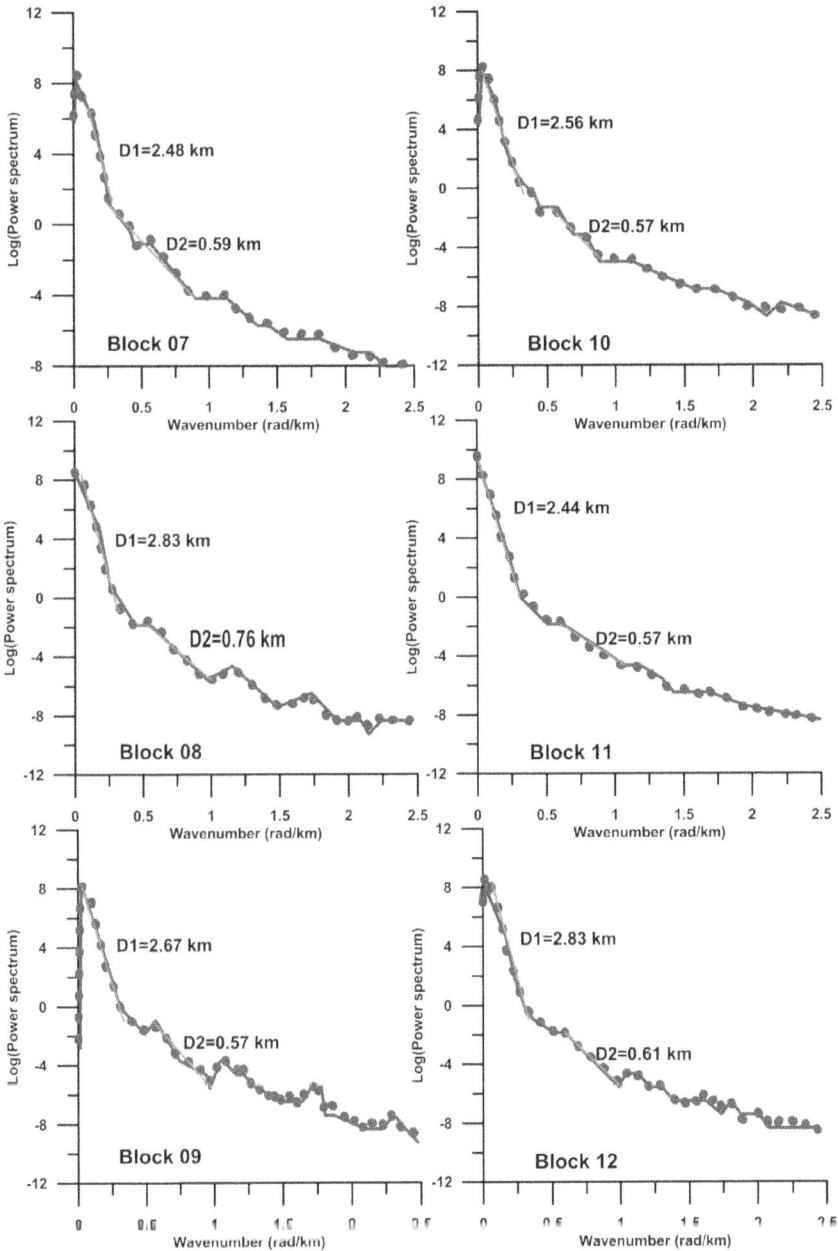

Figure 6.
Power spectra of blocks (B07, B02, ... , B12), showing the depths (D1 & D2) estimated.

blocks labeled B1, B2, ... , B17 for estimations with 50% overlapping are shown. The approach of spectral analysis for the estimation of sedimentary thickness has been done. **Figures 5–7**, showed the seventeen power spectra generated from the gravity anomaly. In the area, basement depth between 2.01 km and 3.73 k is calculated and interpreted the sedimentary basement and shallow sources interpreted in term of depth to top of intrusions in the area between the depths of 0.45 km and 0.76 km (**Figure 8**).

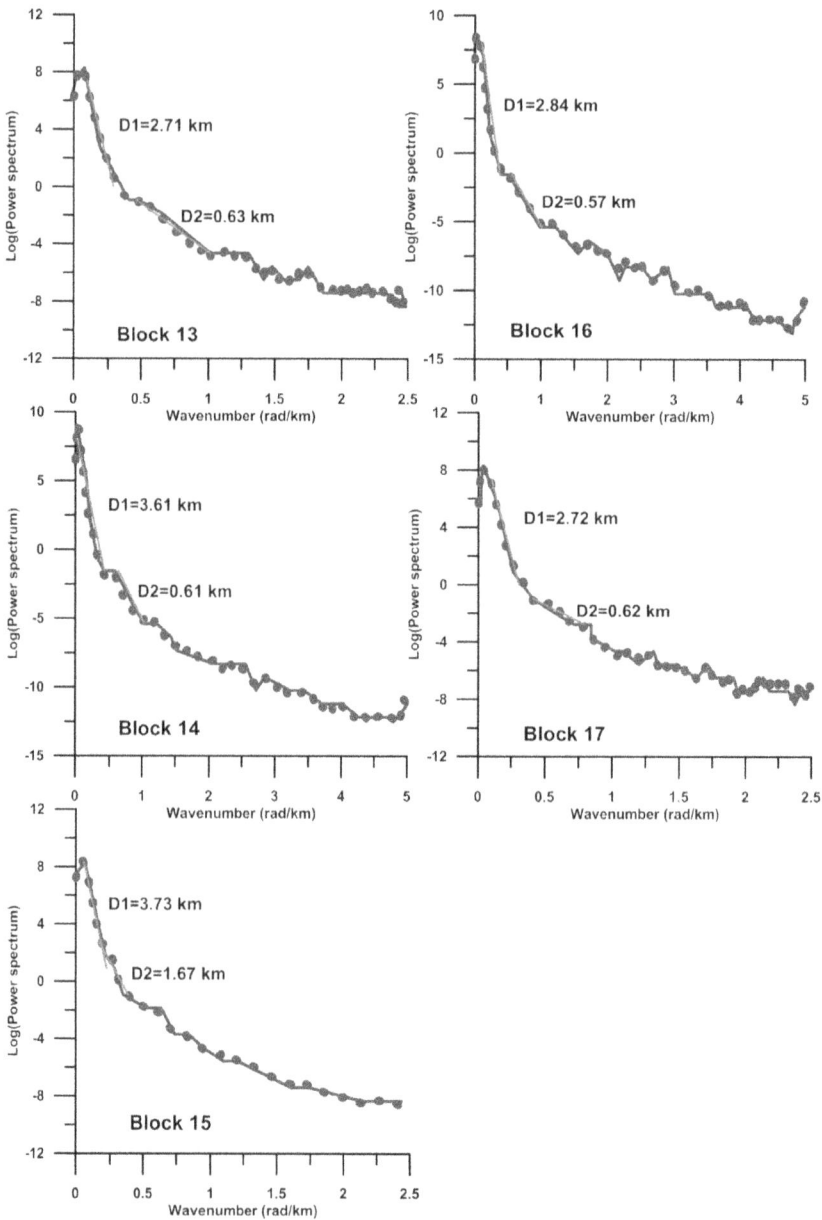

Figure 7.
Power spectra of blocks (B13, B02, ..., B17), showing the depths (D1 & D2) estimated.

4. Discussion and conclusions

Gravity anomalies and depth estimation from airborne gravity data is presented. Spectral analysis approach can be used to design special filters for separation of complete Bouguer anomalies into the residual and regional components. The power spectrum from potential field data for depth interpretation using spectral analysis approach is always not straight due to randomly uncorrelated distributions of anomalous sources. The depth estimation technique using spectral analysis can be

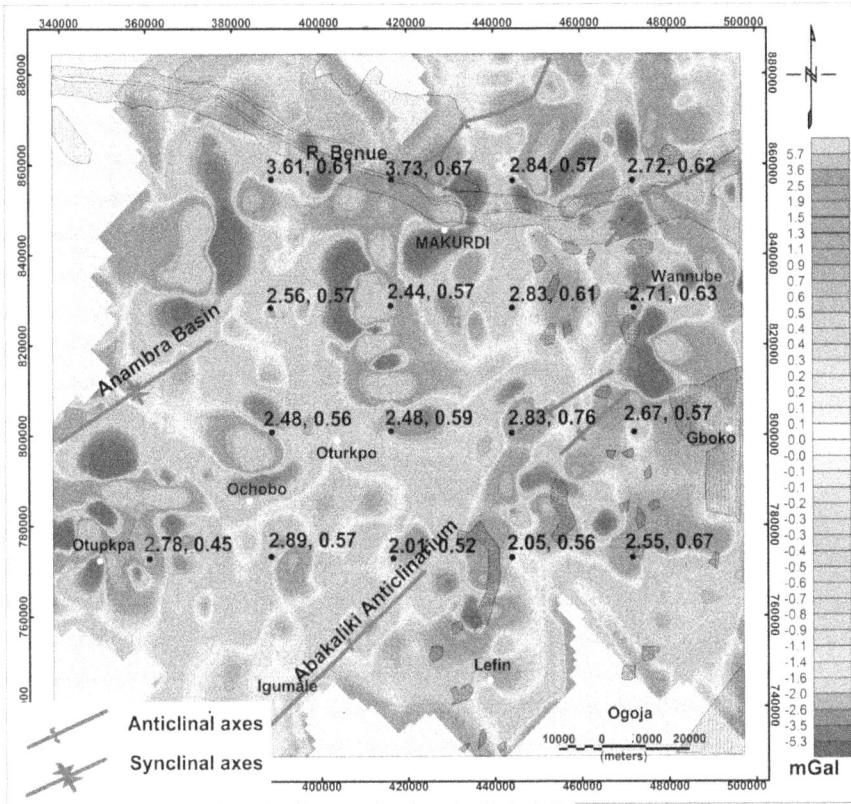

Figure 8.
Residual component of the complete Bouguer gravity anomaly map. Estimated depths (D1 & D2) of each block (B01, B02, …, B17) of 55 km x 55 km with 50% overlap are shown for the sedimentary basement estimation. Gboko as well as the basement outcrops east of Ogoja are shown. Intrusions and locations of the anticline and syncline are superimposed.

applied either in space or frequency domain. However, depth estimation of anomalous sources in frequency/wavenumber domain is simple and more reliable due to the fact that, in wavenumber domain, the convolution operator is conveniently transformed to multiplication notation using Fourier Transform. The estimation of depth from the approach is normally done from the slope of the logarithmic plot of the power spectrum against the wavenumber [12]. Geological and tectonic complexity of a region can be interpreted from the technique of spectral analysis. The drawback in using the approach of spectral analysis for depth interpretation, is that, the concept is based on an ensemble of prisms of frequency-independent randomly and uncorrelated distribution of anomalous sources equivalent to white noise distribution. That is to say, spectral analysis technique involves manually selecting a segment that correspond to certain wavenumber range from the power spectra to represents a detectable anomalous sources inside the earth.

Acknowledgements

I am thankful to geophysical Department of the Nigerian Geological Survey Agency (NGSA), Abuja office, for providing the airborne gravity data used for the present study.

Conflict of interest

I declare no conflict of interest as regards this chapter.

Author details

Mukaila Abdullahi
Modibbo Adama University, Yola, Nigeria

*Address all correspondence to: mukailaa.agp@mautech.edu.ng

IntechOpen

References

[1] Rao, C.V., Chakravarthi, V. and Raju, M.L., 1993. Parabolic Density Function in Sedimentary Basin Modelling. Pure and Applied Geophysics 140(3), 493-501.

[2] Anderson, N.L., Essa, K.S. and Elhussein, M., 2020. A comparison study using particle swarm optimization inversion algorithm for gravity anomaly interpretation due to a 2D verticle fault structure. Journal of Applied Geophysics 179, 104120.

[3] Essa, K.S., Géraud, Y. and Diraison, M., 2021. Fault parameters assessment from the gravity data profiles using the global particle swarm optimization. Journal of Petroleum Science and Engineering 207, 109129.

[4] Essa, K.S., Mehanee, S.A. and Elhussein, M., 2021. Gravity data interpretation by a two-sided fault-like geologic strucyure using the global particle swarm technique. Physics of the Earth and Planetary Interior 311, 106631.

[5] Essa, K.S. and Abo-Ezz, E.R., 2021. Potential field data interpretation to detect the parameters of buried geometries by applying a nonlinear least-squares approach. Acta Geodaetica et Geophysica 56, 387-406.

[6] Rao, C.V., Pramanik, A.G., Kumar, G.V.R.K. and Raju, M.L., 1994. Gravity interpretation of sedimentary basins with hyperbolic density contrast. Geophysical Prospecting 42, 825-839.

[7] Chakravarthi, V., 1995. Gravity Interpretation of Nonoutcropping Sedimentary Basins in Which the Density Contrast Decreases Parabolically with Depth. Pure and Applied Geophysics 145, 327-335.

[8] Chakravarthi, V., Singh, S.B. and Babu, G.A., 2001. INVER2DBASE-A program to compute basement depths of density interfaces above which the density contrast varies with depth. Computer and Geosciences 27, 1127-1133.

[9] Chakravarthi, V. and Sundararajan, N., 2005. Gravity modelling of 21/2 sedimentary basins – a case of variable density contrast. Computer and Geosciences 31, 820-827.

[10] Pallero, J.L.G., Fernandez-Martinez, J.L., Bonvalot, S. and Fudym, O., 2015. Gravity inversion and uncertainty assessment of basement relief via Particle Swarm Optimization. Journal of Applied Geophysics 116, 180-191.

[11] Singh, A. and Biswas, A., 2016. Application of Global Particle Swarm Optimization for Inversion of Residual Gravity Anomalies over Geological Bodies with Idealized Geometries. Natural Resources Research.

[12] Abdullahi, M., Singh, U.K. and Roshan, R., 2019. Mapping magnetic lineaments and subsurface basement beneath parts of Lower Benue Trough (LBT), Nigeria: Insights from integrating gravity, magnetic and geologic data. Journal Earth System Sciences, 128:17.

[13] Bott, M.II.P., 1960. The use of rapid digital computing methods for direct gravity interpretation of sedimentary basins. Geophysical Journal of Royal Astronomical Society 3, 63-67.

[14] Eckhardt, E.A., 1940. A Brief History of the Gravity Method of Prospecting for oil. Geophysics 5(3), 231-242.

[15] Krahenbuhl, R.A. and Li, Y., 2012. Time-lapse gravity: A numerical demonstration using robust inversion and joint interpretation of 4D surface and borehole data. Geophysics 77(2), G33-G43.

[16] Adighije, C., 1981. A gravity interpretation of the Benue Trough, Nigeria. Tectonophysics, 79, 109-128.

[17] Ali, M.Y., Watts, A.B. and Farid, A., 2014. Gravity anomalies of the United Arab Emirates: Implications for basement structures and infra-Cambrian salt distribution. GeoArabia 19(1), 85-112.

[18] Essa, K.S., Mehanee, S.A., Soliman, K.S. and Diab, Z.A., 2020. Gravity profile interpretation using the R-parameter imaging technique with application to ore exploration. Ore Geology Reviews126, 103695.

[19] Ajayi C.O. and Ajakaiye, D.E., 1986. Structures deduced from gravity data in the middle Benue Trough, Nigeria. Journal of African Earth Science 5, 359–369.

[20] Bhattacharyya, B.K., 1965. Two dimensional harmonic analysis as a tool for magnetic interpretation. Geophysics 30, 829-857.

[21] Spector, A. and Grant, F.S., 1970. Statistical model for interpreting aeromagnetic data. Geophysics 35, 293–302.

[22] Maurizio, F., Tatina, Q. and Angelo, S., 1998. Exploration of a lignite bearing in Northern Ireland, using ground magnetic. Geophysics 62, 1143-1150.

[23] Tanaka, A., Okubo, Y. and Matsubayashi, O., 1999. Curie point depth based on spectrum analysis of the magnetic anomaly data in East and Southeast Asia. Tectonophysics 306, 461–470.

Chapter 6

Gravity Anomaly Interpretation Using the R-Parameter Imaging Technique over a Salt Dome

Khalid S. Essa and Zein E. Diab

Abstract

Rapid imaging technique, so-called "R-parameter", utilized for interpreting a gravity anomaly profile. The R-parameter based on calculating the correlation factor between the analytic signal of the real anomaly and the analytic signal of the forward anomaly of assumed buried source denoted by simple geometric shapes. The model parameters (amplitude, origin, depth, and shape factor) picked at the maximum value of the R-parameter. The technique has been proved on noise free and noisy numerical example, numerical example showing the impact of interfering sources. Furthermore, the introduced technique has been successfully applied to visualize a salt dome gravity anomaly profile, USA. The obtained results are in good agreement with those reported in the published studies and that with that obtained from drilling.

Keywords: anomaly, salt dome, R-parameter method, depth

1. Introduction

Gravity data interpretation has been widely used to appraise the different types of subsurface structures and their locations [1–8]. Gravity methods have been widely applied to ore and mineral exploration [9–13], hydrocarbon exploration [14–16], cave detection [17, 18], hydrogeology [19, 20], geothermal and volcanic activity [21–23], locating of unexploded military ordnance [24], environmental and engineering application [25, 26] and archaeological investigations [27, 28].

The quantitative interpretation of gravity data using simple models (spheres and cylinders) is common in exploratory geophysics and continues to be of interest [29–33]. In geologic contexts with a single gravity anomaly, it can be quite appropriate [34]. A single isolated causal body can invert this recorded gravity anomaly to establish its distinctive inverted parameters and fit the recorded data.

The simple geometric models can be matched with the subsurface structures encountered during application of several approaches for inversion [35–39]. These methods include graphical and numerical characteristic points approaches [40–42], ratio technique [43], Fourier transform method [44], the neural network algorithms [45], Mellin transform technique [36], and Werner deconvolution technique [46]. However, the drawbacks of these methods based on tending to generate high number of invalid solution due to few numbers of points and data used, noise or window size incompatibility. As a result, these approaches are subjective, which can lead to significant inaccuracies in calculating the buried anomalous body's character-istic inverse parameters [41, 47], which is to be expected. Gupta [48] and Essa [49] developed techniques depending on successive minimization approaches, which uti-lize the whole measured data to assess the depth parameter and then used some of characteristic points to continue in estimating the rest parameters such as amplitude coefficient. Shaw and Agarwal [37] used the Walsh transform scheme to determine the depth of buried bodies. Mehanee [47] used the regularized conjugate gradient method to construct an effective iterative method based on the use of logarithms of the model parameters for gravity inversion. The method inverts the residual gravity data acquired along profile for evaluating a depth and amplitude coefficient of buried bodies and suitable for subsurface imaging and mineral exploration.

Here, the study proposed an application of the robust R-parameter imaging method to interpret residual gravity data along a profile over idealized geometric bodies such as semi-infinite vertical cylinder, infinitely long horizontal cylinder, and sphere models. The goal is to establish the underlying approximative model by determining the body parameters, which include its origin, depth, amplitude coef-ficient, and shape. The R-parameter imaging method depends on the correlation coefficient amongst the analytic signal of the collected and calculated gravity data. The optimum solution occurs at the maximum R-parameter value.

The benefit behind the use of this method is fall in estimating the depth and body location with an acceptable value compared to the true ones and used the whole gravity data points of the profile, instead of just a few characteristic points. In addition to the method does not require priori information of the subsurface and directly interpret the anomaly from the given observed data. This chapter begins with a layout of the forward modeling, which contains a theoretical gravity formula, an R-parameter imaging approach description, numerical models test without and including noise, and a field data for slat dome investigation.

2. Forward modeling

A closed-form solution for the gravity anomaly caused by simple geometric structures at a measured point (x_j) along a profile (**Figure 1**) is given by [29, 50, 51].

$$g(x_j, x_o, z, z_o, \eta, q, A) = A \frac{(z_o - z)^\eta}{\left[(x_j - x_o)^2 + (z_o - z)^2 \right]^q}, \quad j = 1, 2, 3, \dots, n \quad (1)$$

where x_j and x_o are the X-coordinates of the measured points and the buried body center, z and z_o are the Z-coordinates of the measured point and the subsur-face source (the z axis is chosen positive downward) (**Figure 1**), q is the shape, and A is the amplitude coefficient. The appropriate explanations of A, q, and η for the above-mentioned simple bodies are shown in **Table 1**.

3. Methodology

The gravitational anomaly's analytic signal is written as follows [52, 53]:

$$\xi_s(x) = \frac{\partial g}{\partial x} - i \frac{\partial g}{\partial z}, \quad i = \sqrt{-1}, \quad (2)$$

where $\frac{\partial g}{\partial x}$ and $\frac{\partial g}{\partial z}$ are the horizontal and vertical gradients of the gravity anomaly.

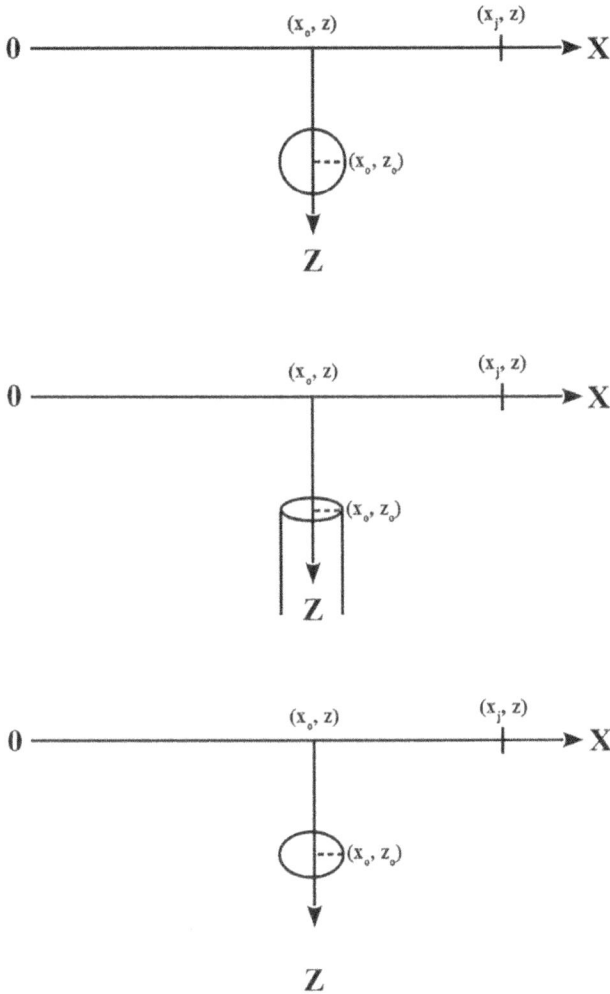

Figure 1.
Geometry and parameters of sphere (top), semi-infinite vertical cylinder (middle and infinitely long horizontal cylinder (bottom) (re-drawn from [39]).

Case	A	η	q
Sphere	$\frac{4}{3}\pi\gamma\sigma r^3$	1	3/2
Horizontal cylinder	$2\pi\gamma\sigma r^2$	1	1
Vertical cylinder	$\pi\gamma\sigma r^2$	0	1/2

Table 1.
Definitions of A, η and q of the simple-geometric bodies.

The amplitude of the analytic signal $|\xi_s(x)|$ can be calculated as follows [44]:

$$|\xi_s(x)| = \sqrt{\left(\frac{\partial g}{\partial x}\right)^2 + \left(\frac{\partial g}{\partial z}\right)^2}.$$

(3)

By an adapting the horizontal and vertical derivatives to Eq. (1), and putting the obtained outcomes into Eq. (3), we get the following:

$$|\xi_s(x)| = A(z_o - z)^{(\eta-1)} \frac{\sqrt{4q^2(x_j - x_o)^2(z_o - z)^2 + \left(\eta(x_j - x_o)^2 - (z_o - z)^2(2q - \eta)\right)^2}}{\left((x_j - x_o)^2 + (z_o - z)^2\right)^{(q+1)}},$$

(4)

where j = 1, 2, 3, ..., n. To calculate the horizontal location (x_o) and depth (z_o) of the buried target (**Figure 1**), the 2-D X-Z mosaic of the correlation coefficient (R) is constructed from the analytic signal amplitudes $|\xi_{so}(x)|$ of the measured data and $|\xi_{st}(x)|$ calculated from the theoretical generated data by a supposed simple-geometric source S(x_o, z_o), and is expressed as:

$$R(x_o, z_o) = \frac{\sum_j^n |\xi_{so}(x)|_j |\xi_{st}(x)|_j}{\sqrt{\sum_j^n |\xi_{so}(x)|_j^2 \sum_j^n |\xi_{st}(x)|_j^2}}.$$

(5)

The analytic signal $|\xi_{so}(x)|$ is assessed numerically along the profile using Eq. (3). To map the relevant discrepancy of the R-parameter from which x_o and z_o are appraised, discretization in the X- and Z-directions is done around the anticipated spatial location of the supposed source. The R-parameters value (R-max) reaches the maximum when the depth and location of the assumed source match the true ones.

4. Numerical examples

To verify constancy in performance of the proposed method, numerical example without noise (noise-free) and with a 20% random noise (noisy) is tested. Another numerical example to evaluate the accuracy and stability in assessing the model parameters in case of interference/neighboring influence.

4.1 Model 1

4.1.1 Noise-free data

The R-parameter imaging method is applied to noise-free numerical gravity anomaly due to simple model consisting of a a horizontal cylinder model (q = 1) with z_o = 5 m, A = 100 mGal m, and x_o = 51 m along a 101-m profile length (**Figure 2a**). This anomaly is the observable (measured) data that needs to be interpreted. The suggested methodology was initiated with estimating the horizontal and vertical derivative anomalies of the residual anomaly in **Figure 2a** (**Figure 2b**), and then calculate the amplitude of the analytic signal (**Figure 2c**). The 2-D mosaic surface S is constructed with depending on calculating the R-parameter values and discretized into 1-m intervals in X- and Z-directions and covers an area of 101 × 11 m in the X- and Z directions. Based on a priori information, the range of the model parameters was selected. The R-parameter values were obtained from (Eq. (5)) by setting the shape to its true value (i.e., q = 1) and utilizing the abovementioned ranges of the model location parameters x_o and z_o. The R-max value is represented by a black dot and equal 1 (**Figure 2d**). This is demonstrating that the methodology fruitfully recovered the real values of the location of the inferred gravity profile's origin point (x_o = 51 m) and the depth of the target (z_o = 5 m).

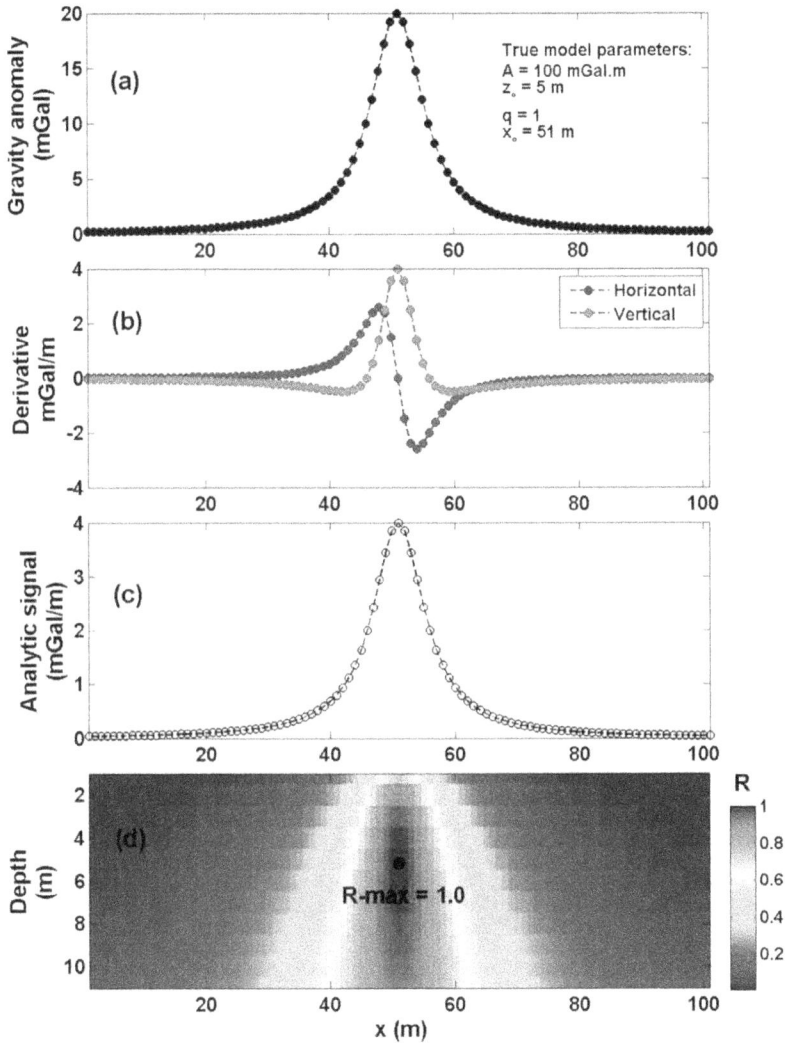

Figure 2.
Model 1: Noise-free data. (a) Horizontal cylinder gravity anomaly, (b) Horizontal and vertical gradients of (a),
(c) Analytic signal anomaly using the data of (b), and (d) 2-D mosaic of the R-parameter and the R-max value.

Table 2 shows a different shape values that employed in the interpretation process. The results (**Figure 3** and **Table 2**) reveal that at q = 1, the R-max = 1 and indicating that the method is stable and capable of capturing the[l] exact values of the model parameters.

We applied the same procedures (by utilizing Eq. (4) as the forward modelling formula in this case) to the analytic signal data presented in **Figure 2c** to explore the recital of the current scheme when used to the analytic signal data themselves instead of the residual gravity data. **Figure 4** shows the outcomes, which are match with those derived from the above-mentioned elucidation of gravity data (**Table 3**).

4.1.2 Noisy data

Given the lack of totally noise-free gravity field data, a 20% random noise (**Figure 5a**) has been introduced to the data in **Figure 2a**. The horizontal and vertical

Shape factor	Maximum R-parameter
(q)	(R-max)
0.5	0.6696
0.6	0.7768
0.7	0.8897
0.8	0.9620
0.9	0.9930
1	**1.0000**
1.1	0.9962
1.2	0.9883
1.3	0.9791
1.4	0.9700
1.5	0.9611

Table 2.
Model 1: Noise-free data. The R-parameter computed for the different shape factors.

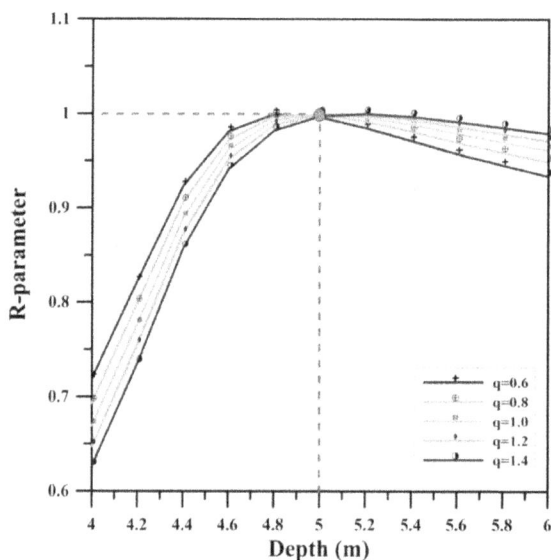

Figure 3.
Model 1: Noise-free data. The R-parameter, depth and shape factor relationship.

derivatives, besides the magnitude of the analytic signal of the measured gravity anomaly, are depicted in **Figure 5b** and **c**. The R-parameter values were evaluated utilizing Eq. (5) and created a 2-D mosaic surface (**Figure 5d**). The maximum R-parameter value is 0.94. The imaging-derived model parameters (z_o = 6 m and x_o = 51 m) are quite near to the real values, indicating that the established technique is stable.

The amplitude coefficient (A) is substantially overstated (**Figure 6a–d**) when the analytic signal data (**Figure 5c**) is interpreted. This is unsurprising given that the technique tries to fit the real data, and the anomalous body's inferred depth (z) is likewise overstated (**Figure 6a**). **Table 4** shows a comparison of model parameters derived by the established technique from the elucidation of each analytic signal and

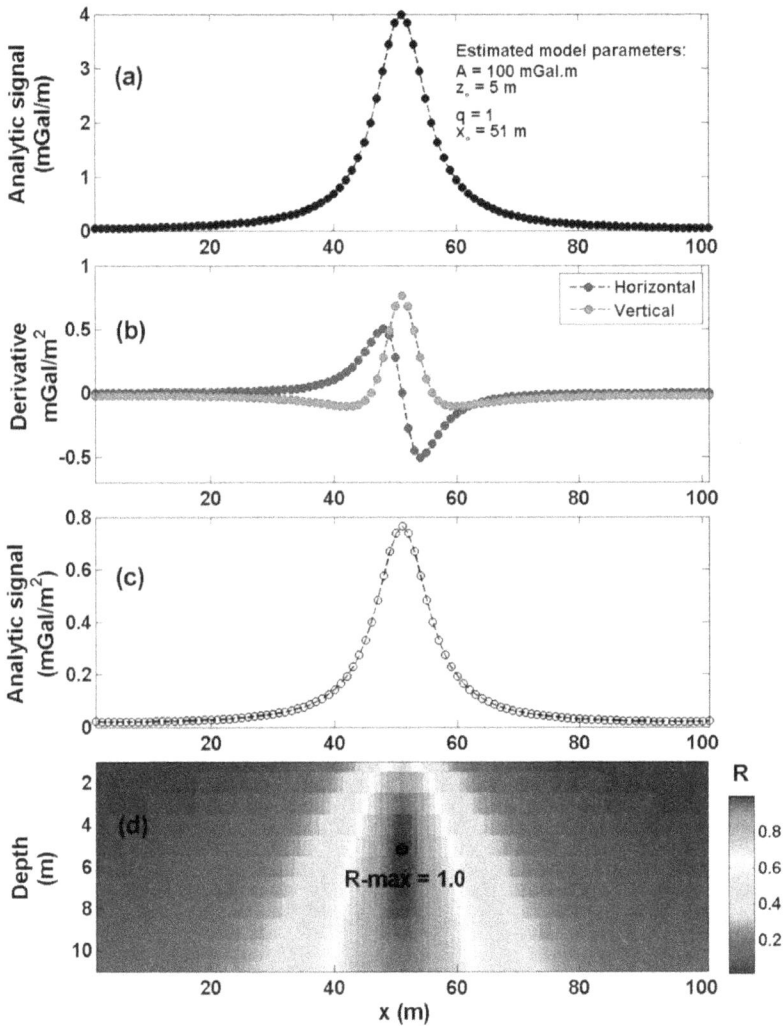

Figure 4.
*Model 1: Noise-free data. (a) Analytic signal anomaly (**Figure 2c**) subjected to the same interpretation, (b) Horizontal and vertical gradients of (a), (c) Analytic signal anomaly using the data of (b), and (d) 2-D mosaic of the R-parameter and the R-max value.*

Estimated model parameters	Analytic signal data	Gravity anomaly data
A (mGal.m)	100	100
z_o (m)	5	5
q	1	1
x_o (m)	51	51

Table 3.
Model 1: Noise-free data. Comparison between the model parameters estimated from the interpretation of using residual anomaly and analytic signal anomaly.

residual anomaly data. This investigation shows that using the given technique to analyses gravity data produces more precise findings than using analytic signal data.

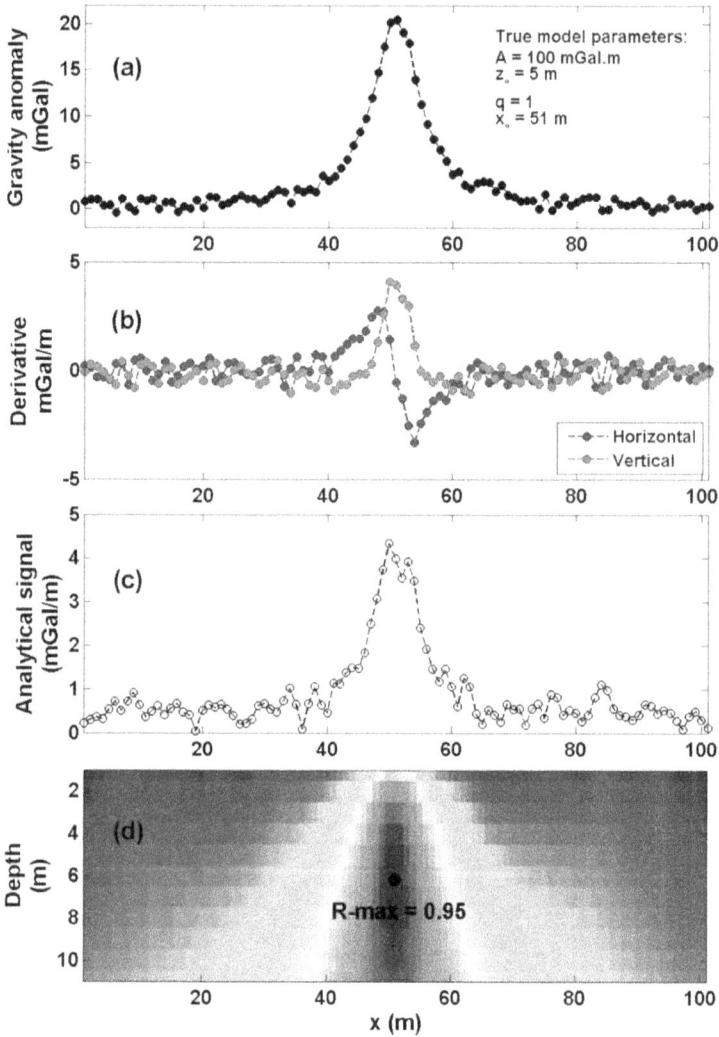

Figure 5.
*Model 1: noisy data. (a) Noisy gravity anomaly of **Figure 2a** after adding 20% noise level, (b) Horizontal and vertical gradients of (a), (c) Analytic signal anomaly using the data of (b), and (d) 2-D mosaic of the R-parameter and the R-max value.*

4.2 Model 2: neighboring effect

The performance of the proposed inversion method with complicated field anomalies and the effect of interfering subsurface structures was investigated. To achieve this, we once again generate a synthetic model data from multiple source bodies as a horizontal cylinder model with $A_1 = 100$ mGal m, $z_1 = 3$ m, $x_1 = 30$ m, and $q_1 = 1$) and a sphere model with $A_2 = 400$ mGal m^2, $z_2 = 4$ m, $x_2 = 80$ m, and $q_2 = 1.5$) (**Figure 7a**).

Figure 7b and **c** illustrate the horizontal and vertical gradients of the composite gravity anomaly, as well as the amplitude of the analytic signal. The R-parameter values were determined using Eq. (5) for each source location and a 2-D mosaic surface S of 101 × 11 m in the X- and Z-directions constructed and discretized into 1-m intervals in both directions. The 2-D mosaic (**Figure 7d**) indicates that the R-max value for each source is 0.8 and 0.62 at q equal 1 and 1.5, respectively.

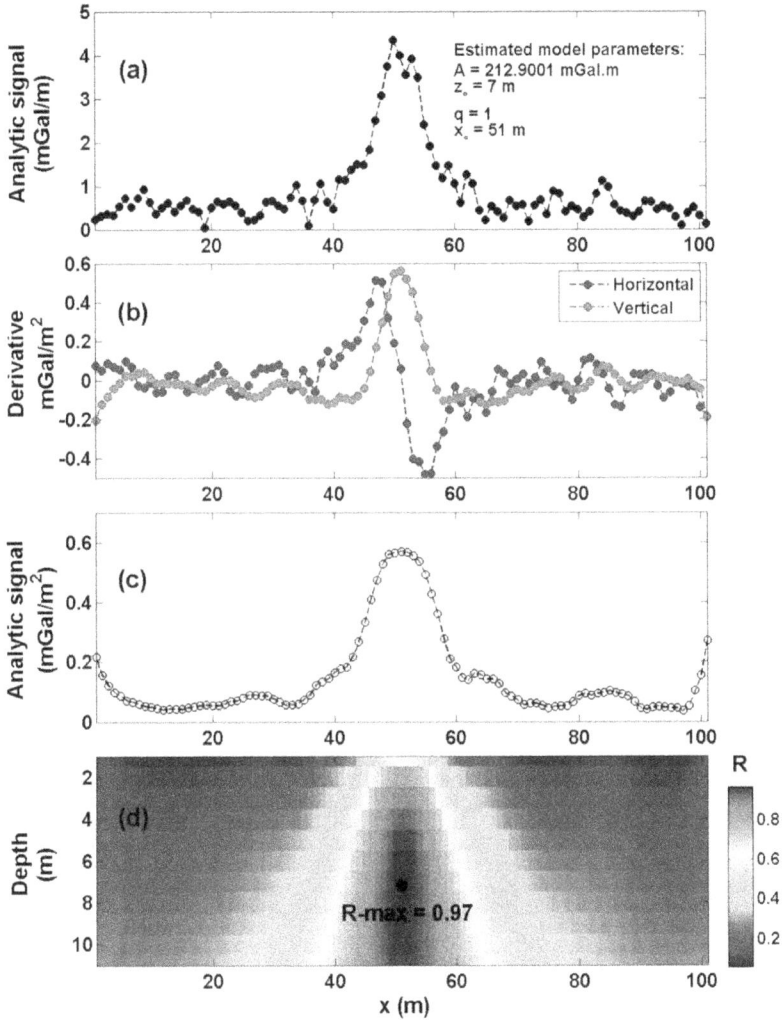

Figure 6.
*Model 1: Noisy data. (a) Analytic signal anomaly (**Figure 5c**) subjected to the same interpretation, (b) Horizontal and vertical gradients of (a), (c) Analytic signal anomaly using the data of (b), and (d) 2-D mosaic of the R-parameter and the R-max value.*

Estimated model parameters	Analytic signal data	Gravity anomaly data
A (mGal m)	212.90	100
z_o (m)	7	5
q	1	1
x_o (m)	51	51

Table 4.
Model 1: Noisy data. Comparison between the model parameters estimated from the interpretation of using residual anomaly and analytic signal anomaly.

The obtained model parameters for horizontal cylinder and sphere are $A_1 = 120.1$ mGal m, $z_1 = 3.6$ m, and $x_1 = 30$ m and $A_2 = 443.1$ mGal m^2, $z_2 = 4.2$ m, and $x_2 = 80$ m, respectively, which the results shows that the method is stable and robust.

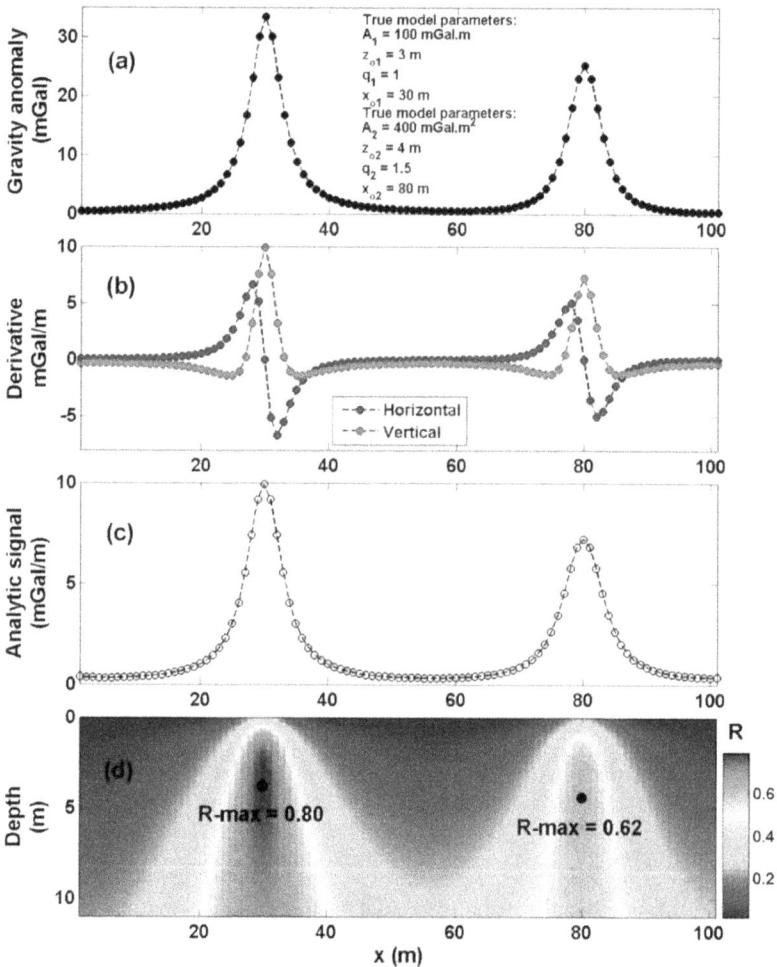

Figure 7.
Model 2: Interference/neighboring effect. (a) Gravity anomaly generated by two different adjacent bodies, (b) Horizontal and vertical gradients of (a), (c) Analytic signal anomaly using the data of (b), and (d) 2-D mosaic of the R-parameter and the R-max values.

To better understand the procedure, we tainted the composite anomaly (**Figure 7a**) with a 20% noise level (**Figure 8a**). The horizontal and vertical derivatives, as well as the corresponding amplitude of the analytic signal, are shown in **Figure 8b** and **c**. The retrieved R-parameter image is shown in **Figure 8d**, with R-max values of 0.79 and 0.61 for a horizontal cylinder and a sphere, respectively. The drop in maximum parameter values compared to (**Figure 7d**) is attributable to the noise introduced into the data as well as the effect of the nearby objects. The model parameters for the first and second bodies revealed by imaging are: $A_1 = 126.9$ mGal m, $z_1 = 3.8$ m, and $x_1 = 30$ m and $A_2 = 556.3$ mGal m^2, $z_2 = 4.7$ m, and $x_2 = 80$ m, respectively, which are quite adjacent to the real values.

Figure 9 depicts the results of the study of the noisy analytic signal data seen in **Figure 8c**. The amplitude coefficients and burial depths recovered from the elucidation are exaggerated (**Figure 9a–d**), as shown in **Table 5**, which coincides with and confirms the aforementioned results.

On the basis of the theoretical models presented above, it can be inferred that the technique described here is stable and robust.

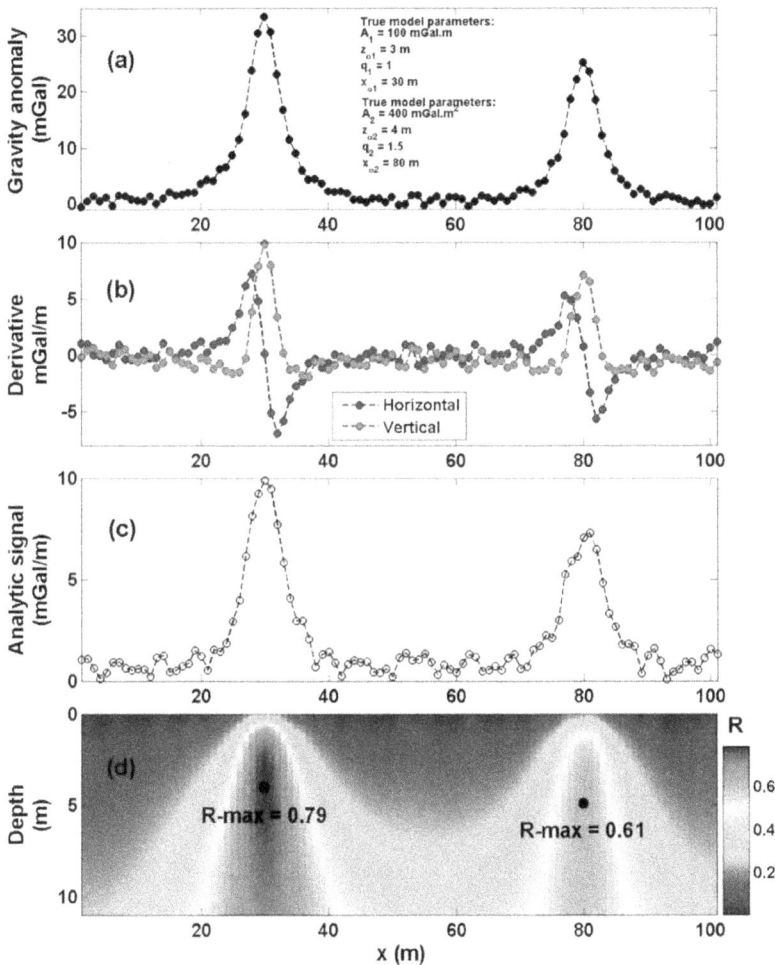

Figure 8.
*Model 2: Interference/neighboring effect with noise. (a) Noisy gravity anomaly of **Figure 7a** after adding 20% noise level, (b) Horizontal and vertical gradients of (a), (c) Analytic signal anomaly using the data of (b), and (d) 2-D mosaic of the R-parameter and the R-max values.*

5. Field data

A published field example over a salt dome anomaly is examined in order to thoroughly test the applicability of the established methodology. For a variety of reasons, this case was chosen. First, the residual gravity profile was created by a simple body that may be truthfully inferred. Second, the drilling information helps in estimating the density contrast of the underlying body. Knowing the density contrast, the radius can be calculated and the depth to the top also can be inferred by using the definition of the amplitude coefficient (**Table 1**). Moreover, the depth of the vertical cylinder model is measured to the top but the depth of a horizontal cylinder and sphere model is measured to the center of the body (**Figure 1**). Third, the gravity data was taken from an area with recognized drilling information, allowing the results obtained from the technique proposed here to be cross-validated against those received via drilling.

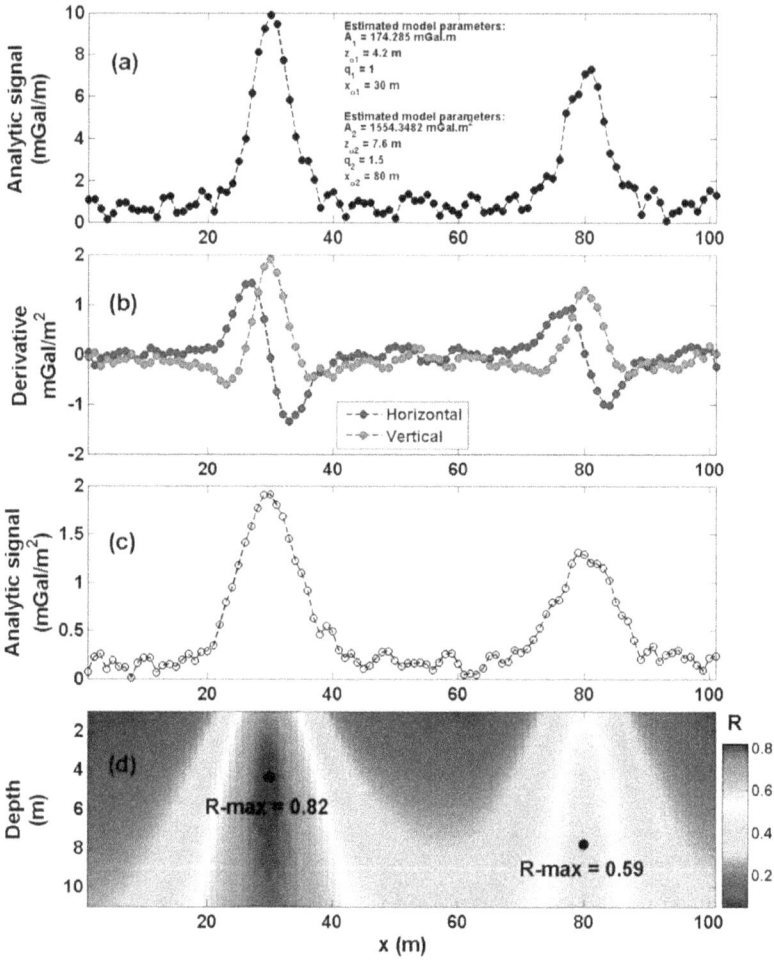

Figure 9.
*Model 2: Interference/neighboring effect with noise. (a) Analytic signal data (**Figure 8c**) subjected to the same interpretation, (b) Horizontal and vertical gradients of (a), (c) Analytic signal anomaly using the data of (b), and (d) 2-D mosaic of the R-parameter and the R-max values.*

Estimated model	Noisy contaminated interference/neighboring effect			
	Analytic signal data		Gravity anomaly data	
Parameters	First anomaly	Second anomaly	First anomaly	Second anomaly
A (mGal m$^{2q-\eta}$)	174.2 mGal m	1554.3 mGal m^2	126.9 mGal m	556.3 mGal m^2
z_o (m)	4.2	7.6	3.8	4.7
q	1	1.5	1	1.5
x_o (m)	30	80	30	80

Table 5.
Model 2: Interference/neighboring effect with noise. Comparison between the model parameters estimated from the interpretation of using residual anomaly and analytic signal anomaly.

5.1 Humble dome anomaly, Houston, USA

Gravity map was acquired over the Humble Dome, Houston, Texas. At the earth's surface, the measurements of this salt dome structure reveal a negative circular contoured Bouguer anomaly ([41], Figures 8–16). A Bouguer gravity profile is taken across the center of the Humble salt dome gravity map in Houston ([41], Figures 8–16). The Bouguer gravity profile was subject to a suitable separation method to remove the regional anomaly and obtain the residual gravity anomaly. The residual gravity anomaly profile of about 26 km long was digitized at an interval of 0.26 km (**Figure 10a**).

The R-parameter method procedures were applied to the residual gravity anomaly profile for the available shape parameters (**Table 6**). **Figure 10b–d** express the horizontal and vertical derivative anomalies, the amplitude analytic signal, and the R-parameter 2-D mosaic. It is found that the R-max value is 0.99 corresponds to a

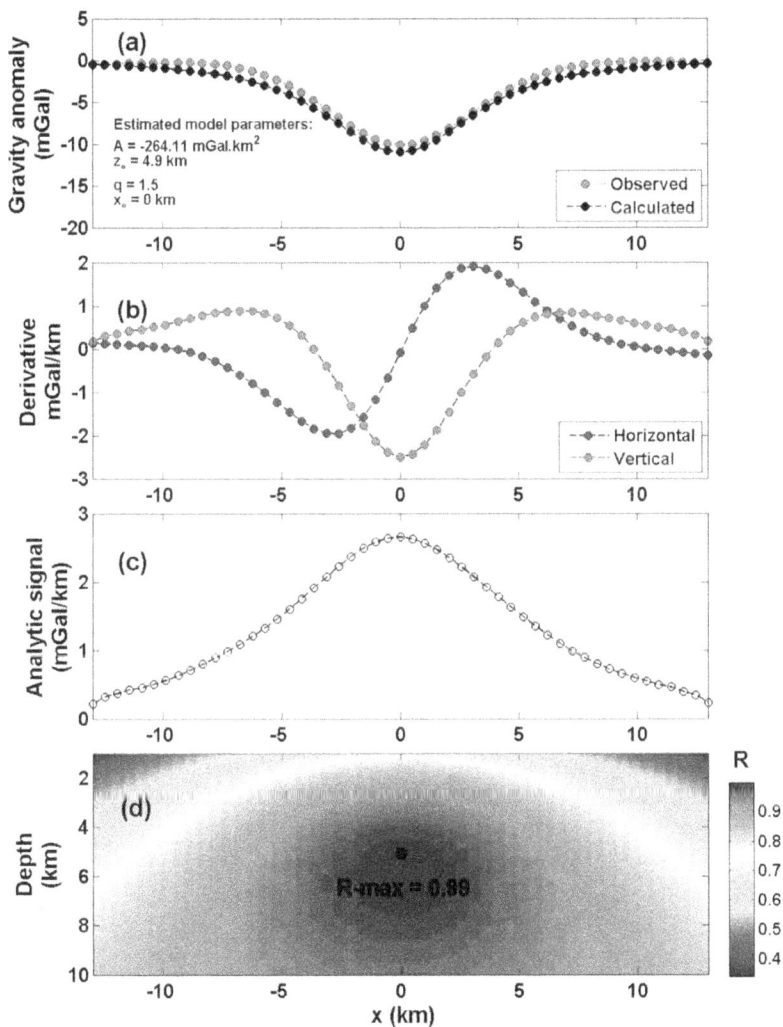

Figure 10.
The Humble dome anomaly, USA. (a) Gravity anomaly profile (red dotted lines) and the optimum-fitting model (solid black line), (b) Horizontal and vertical gradients of (a), (c) Analytic signal anomaly using the data of (b), and (d) 2-D mosaic of the R-parameter and the R-max value.

q	R-max
0.5	0.999462
0.6	0.997341
0.7	0.996274
0.8	0.992509
0.9	0.986895
1	0.984110
1.1	0.987549
1.2	0.992912
1.3	0.996863
1.4	0.998792
1.5	0.999501

Table 6.
The Humble dome gravity anomaly, USA. The R-parameter calculated from different shape factors.

Model parameters	Approaches and techniques of					Present study
	[28]	[33]	[39]	[46]	[47]	
A mGal km^2	—	—	−292.54	—	−279.81	**−269.39**
z_o (km)	4.96	4.97	4.62	4.81	4.58	**4.90**
q	1.5	1.5	1.5	1.5	1.5	**1.5**
x_o (km)	—	—	—	—	—	**0**

Table 7.
The Humble dome gravity anomaly, USA. The Estimated parameters.

spherical shape (q = 1.5). The best estimated model parameters are: q = 1.5, z_o = 4.90 m, x_o = 0 m, and A = −269.39 mGal m^2.

The humble dome anomaly has been interpreted by several authors assuming a spherical source to decide the depth of the salt body. The obtained results agree well with those depths to the center that obtained by the published literatures of [36, 41, 47, 54–56] (**Table 7**).

By using a density contrast of −0.13 gm/cm^3 of [41], then the depth to the top of the spherical body of the humble dome obtained from the proposed technique is 315 m, which in excellent covenant with the true depth (305 m) confirmed by drilling and seismic information [41]. **Table 8** shows that several other researchers

Approaches and techniques of	Depth to the top (m)
[33]	426
[39]	299
[46]	326
[47]	326
Present study	315

Table 8.
Depth to the top of the spherical body of the Humble dome anomaly, obtained by various approaches using an assumed density contrast ($\Delta\rho$) of −0.13 gm/cm^3 [41].

utilizing the same density contrast found some differences in the depths to the top of this spherical source. The use of simple geometrically bodies in the constrained class of spheres, horizontal cylinders, and vertical cylinders is thus suggested as a way to accurately apply the current methodology to extract depth information. As a result, if exact density contrasts are used, the related radii can be correctly computed as well.

5.2 Louisiana dome anomaly, USA

A residual gravity map was acquired over a salt dome off the coast of Louisiana, USA ([41], Figures 8–20). The residual gravity anomaly profile [57] is redrawn across the center of the map, normal to the causal anomaly's striking. The residual gravity anomaly profile of about 13,000 m long was digitized at sampling interval of 200 m (**Figure 11a**).

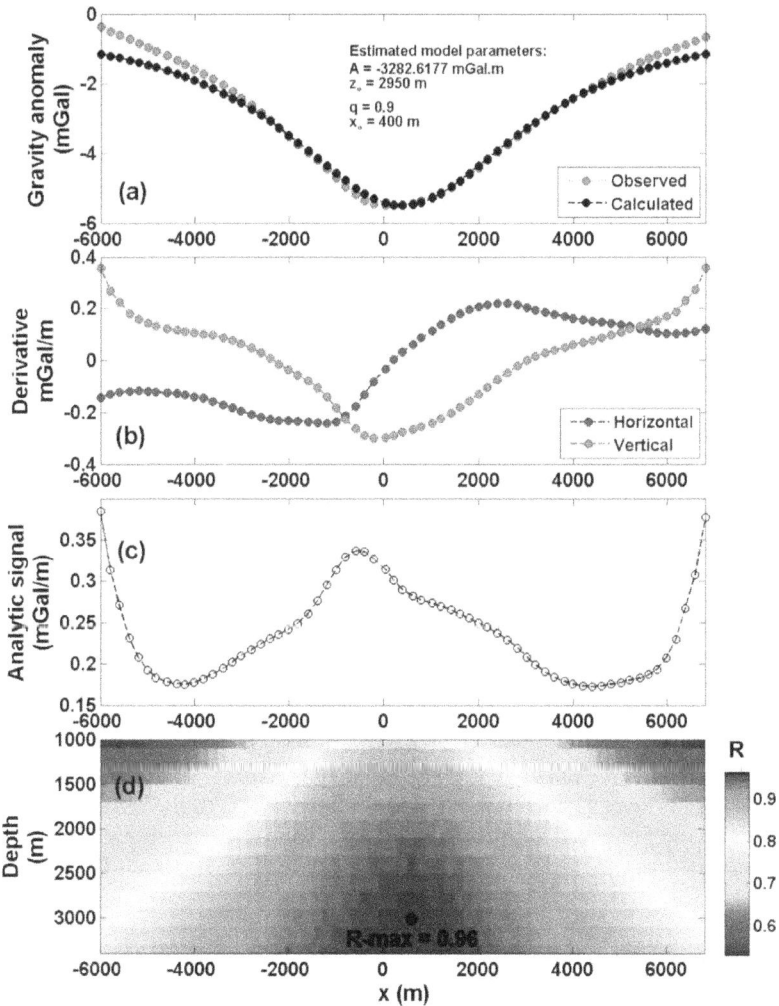

Figure 11.
The Louisiana dome anomaly, USA. (a) Gravity anomaly profile (red dotted lines) and the optimum-fitting model (solid black line), (b) Horizontal and vertical gradients of (a), (c) Analytic signal anomaly using the data of (b), and (d) 2-D mosaic of the R-parameter and the R-max value.

q	R-max
0.5	0.943033
0.6	0.934839
0.7	0.926284
0.8	0.957334
0.9	**0.963889**
1	0.957410
1.1	0.942955
1.2	0.926284
1.3	0.909983
1.4	0.894917
1.5	0.881254

Table 9.
The Louisiana dome gravity anomaly, USA. The R-parameter calculated from different shape factors.

Model parameters	Approaches and techniques of		Present study
	[47]	[5]	
A mGal m	−16400	−16021	−3282.61
z_o (m)	2899	2702.2	2950.00
q	1	1	0.9
x_o (m)	—	506.5	400
Misfit	12.4%	2.9×10^{-3}	1.3×10^{-3}

Table 10.
The Louisiana dome gravity anomaly, USA. The estimated parameters.

By applying the R-parameter method procedures mentioned before to the residual gravity anomaly profile of Louisiana we get the available shape parameters corresponding to the maximum R-parameter (R-max) as shown in **Table 9**. **Figure 11b–d** shows the horizontal and vertical derivative anomalies, the amplitude analytic signal, and the R-parameter 2-D mosaic of the Louisiana anomaly. It is found that the R-max value is 0.96 corresponds to $q = 0.9$ which is approximated by horizontal cylinder shape. The best estimated model parameters are: $q = 0.9$, $z_o = 2950$ m, $x_o = 400$ m, and $A = -3282.61$ mGal m.

The Louisiana dome anomaly has been interpreted by different authors assuming a horizontal source to determine the depth to the center of the salt body. The obtained results have a good agreement with those depths to the center that obtained by the published literatures of [5, 41] (**Table 10**). In addition, the proposed method has the lowest misfit compared to the other method (**Table 10**).

6. Discussion

The proposed method of R-parameter imaging technique was deployed to visualizes the salt dome anomalies from the gravity data measured across a 2D profile. The method fitting the anomaly of the measured gravity profile by a single geometric shape body (sphere & cylinder). Such as the spherical source ($q = 1.5$)

suggestion based on the maximum R-parameter (R-max = 0.99) of the Humble dome anomaly (**Figure 10**) and the approximated horizontal cylinder (q = 0.9) with maximum R-parameter (R-max = 0.96) of the Louisiana dome anomaly (**Figure 11**).

The obtained results by the R-parameter method of the Humble dome anomaly was compared with other results in the published literature (**Table 7**) and confirmed with drilling to insure the depth to the top of the buried anomaly (**Table 8**). For the Louisiana dome anomaly, the obtained results by R-parameter approach was compared with the pervious published literature and weighted by the misfit error between the observed and calculated anomaly for the different techniques used (**Table 10**) to increase the efficiency of the proposed method.

In over all the obtained results using the R-parameter method to investigate the salt dome anomalies is good and acceptable in the two given field examples.

7. Conclusions

In this study, we have introduced and investigated the applicability and the performance of the R-parameter imaging method in elucidating distinctive physical parameters (A, z_o, x_o, q) of simple geometrically-shaped geologic structures (spheres, horizontal cylinders and vertical cylinders) from gravity data along profiles. This inversion imaging method has been demonstrated successfully on numerically generated gravity anomalies corrupted with random noise, applied to cases of anomalies from multiple and interfering structures and finally experimented on a two different field cases for salt domes in USA. The approach presented here can be used to investigate salt domes and perform reconnaissance geological studies.

Author details

Khalid S. Essa* and Zein E. Diab
Geophysics Department, Faculty of Science, Cairo University, Giza, Egypt

*Address all correspondence to: khalid_sa_essa@cu.edu.eg
and khalid_sa_essa@yahoo.com

IntechOpen

References

[1] Paterson NR, Reeves CV. Applications of gravity and magnetic surveys: The state of-the-art in 1985. Geophysics. 1985;**50**:2558-2594

[2] Ateya IL, Takemoto S. Gravity inversion modeling across a 2-D dike like structure—A case study. Earth Planets Space. 2002;**54**:791-796

[3] Essa KS, Di Risio M, Celli D, Pasquali D, editors. Geophysics and Ocean Waves Studies. Rijeka: InTech; 2021. 188 p. DOI: 10.5772/intechopen.87807

[4] Hector B, Séguis L, Hinderer J, Descloitres M, Vouillamoz J, Wubda M, et al. Gravity effect of water storage changes in a weathered hard-rock aquifer in West Africa: Results from joint absolute gravity, hydrological monitoring and geophysical prospection. Geophysical Journal International. 2013;**194**:737-750

[5] Biswas A. Interpretation of residual gravity anomaly caused by a simple shaped body using very fast simulated annealing global optimization. Geoscience Frontiers. 2015;**6**:875-893

[6] Essa KS. A fast interpretation method for inverse modeling of residual gravity anomalies caused by simple geometry. Journal of Geological Research. 2012; **2012**:327037. DOI: 10.1155/2012/327037

[7] Nishijma J, Naritomib J. Interpretation of gravity data to delineate underground structure in the Beppu geothermal field, central Kyushu, Japan, Regional Studies. Journal of Hydrology. 2017;**11**:84-95

[8] Anderson NL, Essa KS, Elhussein M. A comparison study using particle swarm optimization inversion algorithm for gravity anomaly interpretation due to a 2D vertical fault structure. Journal of Applied Geophysics. 2020;**179**:104120

[9] Mehanee S, Essa KS. 2.5D regularized inversion for the interpretation of residual gravity data by a dipping thin sheet: Numerical examples and case studies with an insight on sensitivity and non-uniqueness. Earth, Planets and Space. 2015;**67**:130

[10] Chen YQ, Zhang LN, Zhao. Application of bi-dimensional empirical mode decomposition (BEMD) modeling for extracting gravity anomaly indicating the ore controlling geological architectures and granites in the Gejiu tin-copper polymetallic ore field, southwestern China. Ore Geology Reviews. 2017;**88**:832-840

[11] Essa KS, Munschy M. Gravity data interpretation using the particle swarm optimization method with application to mineral exploration. Journal of Earth System Science. 2019;**128**:123

[12] Essa KS, editor. Minerals. Vol. 142. Rijeka: InTech; 2019. DOI: 10.5772/intechopen.74902

[13] Mehanee SA. Simultaneous joint inversion of gravity and self-potential data measured along profile: Theory, numerical examples, and a case study from mineral exploration with cross validation from electromagnetic data. IEEE Transactions on Geoscience and Remote Sensing. 2022;**60**:1-20. Art no. 4701620. DOI: 10.1109/TGRS.2021.3071973

[14] Yuan B, Song L, Han L, An S, Zhang C. Gravity and magnetic field characteristics and hydrocarbon prospects of the Tobago Basin. Geophysical Prospecting. 2018;**66**: 1586-1601

[15] Essa KS, Gèraud Y. Parameters estimation from the gravity anomaly caused by the two-dimensional horizontal thin sheet applying the global particle swarm algorithm. Journal

Petroleum Science and Engineering. 2020;**193**:107421

[16] Coelho ACQM, Menezes PTL, Mane MA. Gravity data as a faulting assessment tool for unconventional reservoirs regional exploration: The Sergipe–Alagoas Basin example. 2021; **94**:104077

[17] Martínez-Moreno FJ, Galindo-Zaldívar J, Pedrera A, González-Castillo L, Ruano P, Calaforra JM, et al. Detecting gypsum caves with microgravity and ERT under soil water content variations (Sorbas, SE Spain). Engineering Geology. 2015;**193**:38-48

[18] Braitenberg C, Sampietro D, Pivetta T, Zuliani D, Barbagallo A, Fabris P, et al. Gravity for detecting caves: Airborne and terrestrial simulations based on a comprehensive karstic cave benchmark. Pure and Applied Geophysics. 2016;**173**: 1243-1264

[19] Rodell M, Chen J, Kato H, Famiglietti JS, Nigro J, Wilson CR. Estimating groundwater storage changes in the Mississippi River basin (USA) using GRACE. Hydrogeology Journal. 2007;**15**:1

[20] Epuh EE, Okolie CJ, Daramola OE, Ogunlade FS, Oyatayo FJ, Akinnusi SA, et al. An integrated lineament extraction from satellite imagery and gravity anomaly maps for groundwater exploration in the Gongola Basin. Remote Sensing Applications: Society and Environment. 2020;**20**:100346

[21] Lichoro CM, Árnason K, Cumming W. Joint interpretation of gravity and resistivity data from the Northern Kenya volcanic rift zone: Structural and geothermal significance. Geothermics. 2019;**77**:139-150

[22] Njeudjang K, Essi JMA, Kana JD, Teikeu WA, Nouck PN, Djongyang N, et al. Gravity investigation of the

Cameroon Volcanic Line in Adamawa region: Geothermal features and structural control. Journal of African Earth Sciences. 2020;**165**:103809

[23] Mulugeta BD, Fujimitsu Y, Nishijima J, Saibi H. Interpretation of gravity data to delineate the subsurface structures and reservoir geometry of the Aluto-Langano geothermal field, Ethiopia. Geothermics. 2021;**94**:102093

[24] Butler DK, Wolfe PJ, Hansen RO. Analytical modeling of magnetic and gravity signatures of unexploded ordnance. Journal of Environmental and Engineering Geophysics. 2001;**6**(1): 33-46

[25] Tinivella U, Giustiniani M, Cassiani G. Geophysical methods for environmental studies. International Journal of Geophysics. 2013;**2**:950353. DOI: 10.1155/2013/95035

[26] Setyawan A, Fukuda Y, Nishijima J, Kazama T. Detecting land subsidence using gravity method in Jakarta and Bandung Area, Indonesia. Procedia Environmental Sciences. 2015;**23**:17-26

[27] Padín J, Martín A, Anquela AB. Archaeological microgravimetric prospection inside don church (Valencia, Spain). Journal of Archaeological Science. 2012;**39**:2

[28] Klokočník J, Cílek V, Kostelecký J, Bezděk A. Gravity aspects from recent Earth gravity model EIGEN 6C4 for geoscience and archaeology in Sahara, Egypt. Journal of African Earth Sciences. 2020;**168**:103867

[29] Salem A, Ravat D, Mushayandebvu MF, Ushijima K. Linearized least-squares method for interpretation of potential-field data from sources of simple geometry. Geophysics. 2004;**69**(3):783-788

[30] Su Y, Cheng LZ, Chouteau M, Wang XB, Zhao GX. New improved

formulas for calculating gravity anomalies based on a cylinder model. Journal of Applied Geophysics. 2012;**86**: 36-43

[31] Essa KS. Gravity interpretation of dipping faults using the variance analysis method. Journal of Geophysics and Engineering. 2013;**10**:015003

[32] Singh A, Biswas A. Application of global particle swarm optimization for inversion of residual gravity anomalies over geological bodies with idealized geometries. Natural Resources Research. 2016;**25**:297-314

[33] Essa KS, Elhussein M. Gravity data interpretation using new algorithms: A comparative study. In: Zouaghi T, editor. Gravity-Geoscience Applications, Industrial Technology and Quantum Aspect. Rijeka: InTech; 2018. pp. 3-17. DOI: 10.5772/intechopen.68576

[34] Abdelrahman EM, El-Araby TM, Essa KS. Shape and depth solutions from third moving average residual gravity anomalies using the window curves method. Kuwait Journal of Science and Engineering. 2003;**30**:95-108

[35] Nettleton LL. Gravity and magnetics for geologists and seismologists. AAPG Bulletin. 1962;**46**:1815-1838

[36] Mohan NL, Anandababu L, Roa S. Gravity interpretation using the Melin transform. Geophysics. 1986;**51**:14-22

[37] Shaw RK, Agarwal NP. The application of Walsh transform to interpret gravity anomalies due to some simple geometrically shaped causative sources: A feasibility study. Geophysics. 1990;**55**:843-850

[38] Abdelrahman EM, Abo-Ezz ER, Essa KS, El-Araby TM, Soliman KS. A least-squares variance analysis method for shape and depth estimation from gravity data. Journal of Geophysical Engineering. 2006;**3**:143-153

[39] Asfahani J, Tlas M. Estimation of gravity parameters related to simple geometrical structures by developing an approach based on deconvolution and linear optimization techniques. Pure and Applied Geophysics. 2015;**172**:2891-2899

[40] Siegel HO, Winkler HA, Boniwell JB. Discovery of the Mobrun Copper Ltd. sulphide deposit, Noranda Mining District, Quebec. In: Methods and Case Histories in Mining Geophysics: Commonwealth Mining Metallurgical Congress. 6th ed. Vancouver; 1957. pp. 237-245

[41] Nettleton LL. Gravity and Magnetics in Oil Prospecting. New York: McGraw-Hill Book Co.; 1976

[42] Reynolds JM. An Introduction to Applied and Environmental Geophysics. New York: John Wiley and Sons; 1997. 796 p

[43] Bowin C, Scheer E, Smith W. Depth estimates from ratios of gravity, geoid and gravity gradient anomalies. Geophysics. 1986;**51**:123-136

[44] Odegard ME, Berg JW. Gravity interpretation using the Fourier integral. Geophysics. 1965;**30**:424-438

[45] Abedi M, Afshar A, Ardestani VE, Norouzi GH, Lucas C. Application of various methods for 2D inverse modeling of residual gravity anomalies. Acta Geophysica. 2009;**58**:317-336

[46] Kilty KT. Short Note: Werner deconvolution of profile potential field data. Geophysics. 1983;**48**:234-237

[47] Mehanee SA. Accurate and efficient regularised inversion approach for the interpretation of isolated gravity anomalies. Pure and Applied Geophysics. 2014;**171**:1897-1937

[48] Gupta OP. A least-squares approach to depth determination from gravity data. Geophysics. 1983;**48**:360-375

[49] Essa KS. A new algorithm for gravity or self-potential data interpretation. Journal of Geophysics and Engineering. 2011;**8**:434-446

[50] Essa KS. New fast least-squares algorithm for estimating the best-fitting parameters of some geometric-structures to measured gravity anomalies. Journal of Advanced Research. 2014;**5**:57-65

[51] Asfahani J, Tlas M. An automatic method of direct interpretation of residual gravity anomaly profiles due to spheres and cylinders. Pure and Applied Geophysics. 2008;**165**:981-994

[52] Nabighian MN. The analytic signal of two-dimensional magnetic bodies with polygonal cross-section: Its properties and use for automated anomaly interpretation. Geophysics. 1972;**37**:507-517

[53] Klingele EE, Marson I, Kahle HG. Automatic interpretation of gravity gradiometric data in two dimensions: Vertical gradients. Geophysical Prospecting. 1991;**39**:407-434

[54] Essa KS. Gravity data interpretation using the s-curves method. Journal of Geophysics and Engineering. 2007;**4**: 204-213

[55] Asfahani J, Tlas M. Fair function minimization for direct interpretation of residual gravity anomaly pro_les due to spheres and cylinders. Pure and Applied Geophysics. 2012;**169**:157-165

[56] Tlas M, Asfahani J, Karmeh H. A versatile nonlinear inversion to interpret gravity anomaly caused by a simple geometrical structure. Pure and Applied Geophysics. 2005;**162**:2557-2571

[57] Roy L, Agarwal BNP, Shaw RK. A new concept in Euler deconvolution of isolated gravity anomalies. Geophysical Prospecting. 2000;**48**:559-575

New Semi-Inversion Method of Bouguer Gravity Anomalies Separation

D.M. Abdelfattah

Abstract

The workers and researchers in the field of gravity exploration methods, always dream that it is possible one day, to be able to separate completely the Bouguer gravity anomalies and trace rock' formations, and their densities distribution from a prior known control points (borehole) to any extended distance in the direction of the profile lines-it seems that day become will soon a tangible true! and it becomes possible for gravity interpretation methods to mimic to some extent the 2D seismic interpretation methods. Where, the present chapter is dealing a newly 2D semi-inversion, fast, and easily applicable gravitational technique, based on Bouguer gravity anomaly data. It now becomes possible through, Excel software, Matlab's code, and a simple algorithm; separating the Bouguer anomaly into its corresponding rock' formations causative sources, as well as, estimating and tracing its thicknesses (or depths) of sedimentary formations relative to the underlying basement's structure rocks for any sedimentary basin, through using of profile(s) line(s) and previously known control points. The newly proposed method has been assessed, examine, and applied for two field cases, Abu Roash Dome Area, southwest Cairo, Egypt, and Humble Salt Dome, USA. The method has demonstrated to some extent comparable results with prior known information, for drilled boreholes.

Keywords: Bouguer, sedimentary cover, slab, relatively thicknesses, basement

1. Introduction

The measurements and analysis of the variation in gravity over the Earth's surface have become powerful techniques in the investigation of the subsurface structures at various depths [1]. Where the gravity anomaly is often attributed to the lateral variation in density-contrast and therefore, one of its major applications, being is used as a reconnaissance tool for and mapping the basement rock's morphology, and its depth below the sedimentary covering of basins. The most challenging problem of ambiguity, for interpreting the potential-field data (gravity and/or magnetic), is still facing the researchers, where the modeling of potential-field data is a non-linear problem. In general, the reference body or source body (i.e., causative body) is imported into the potential model (gravity and/or magnetic), as the initial approximation of the anomaly source, and its parameters are obtained from available geological and geophysical information [2]. Ambiguity in gravity interpretation is inevitable because of the fundamental incompleteness of real observations; it is, however, possible to provide rigorous limits on possible solutions even with incomplete data [3]. Since a unique solution cannot, in general, be recovered from a set of field measurements, geophysical interpretation is concerned either to determine properties of the subsurface that all possible solutions share or to introduce assumptions to restrict the number of admissible solutions [4].

However, a unique solution may be found, when assigning a simple geometrical shape to the causative body [5]. Also, a unique solution can be found by an attempt for treating the problem of ambiguity with a new vision for analysis of the corrected acquired data (measurements) and related it analytically, logically, or mathematically, to its causative sources, as an attempt of the present research.

The newly proposed method is an attempt to reveal and trace the concealed subsurface geological formations' thicknesses and basement depth at each point of the profile (s), of the Bouguer gravity anomaly map, relatively to the formations' thicknesses and basement' rock depth of a prior known in controlling point (e.g., borehole data). Fortunately, almost most of the geological structures can be approximated, by one or more of the available simple geometrical shape models, to represent the causative sources for gravity anomalies. There are several gravity forward techniques to estimate the depth to basement based on rather different approaches that have been proposed before by many authors, such as [6–9]. The forward modeling of mass distribution is a powerful tool to visualize Free Air and Bouguer gravity anomalies that result from different geological situations [10].

2. The methodology

There is a known fact that any depositions of formations layers were deposited in a basin and were may or not subjected to tectonic, hiatus non-deposition, and/or erosions. Therefore, simply the proposed method is mainly based on that fact to calculate average vertical densities-contrasts for the formation's layers and the basement rock in that basin, whatever the geological setting of such formation's layers.

The method is a direct technique for interpreting the Bouguer gravity anomaly in form of profile (s), to calculate the formations' thicknesses, and formations' depths of the sediments relatively to the depth of basement rocks, where the deposited rock' formations are treated as the Bouguer Horizontal Slabs (BHS) or Infinite Horizontal Slabs (IHS), which are vertically stacked in columns and does not rely on the homogeneity or inhomogeneity of densities' distributions, but only on the average vertical densities-contrasts between each of rock' formations slab's in columns and the basement rocks.

Since the Bouguer gravity anomaly correlates with the lateral variation of density of the crustal rocks, a positive or a negative anomaly is created, whenever there is a change in rock density [11]. And also, the Bouguer anomaly is defined upon the datum level of gravity reduction of an arbitrary elevation [12]. Where this correction is taken into account the attraction of masses between a reference elevation often the sea level, and each of an individually measuring stations' points. In other words, the variation of the Bouguer anomaly should reflect the lateral variation in density, such that a high-density feature in a lower-density medium should give rise to a positive Bouguer anomaly. Conversely, a low-density feature in a higher-density medium should result in a negative Bouguer anomaly [13].

2.1 Infinite horizontal slab equation (IHSE)

The gravity effect or the Bouguer gravity anomaly can be calculated at any point of Cartesian coordinates (x, z) on the surface of the earth or reference measured datum, due to BHS or IHS is given by the Eq. (1) as follows:

$$gB(x, z) = 2\pi\rho hG \tag{1}$$

where gB: is the gravity effect or Bouguer gravity anomaly due to slab in m. Gal (10^{-5} m/s^2).

ρ: is the density of the horizontal slab in gm/cm^3 (10^3 kg/m^3).

h: is the thickness of the IHS in km (10^3 m).

G: is the Universal Gravitational Constant (6.67×10^{-11} Nm^2/kg^2), N: is referring to Newton or force unit.

The Eq. (1) can be rewritten in a modified form for the new method purpose as follows:

$$gB(x, z) = 2\pi G \sum_{i=1}^{N} \overline{\Delta\rho}(i)\Delta z(i) \tag{2}$$

where $\overline{\Delta\rho}$: is the average vertical density-contrast in gm/cm^3 (10^3 kg/m^3).

Δz: is the difference between the depth of top and bottom the IHS in km (10^3 m).

i: is the index number (i = 1, 2, 3 ... N).

N: is the number of rocks 'formations, and

$$\overline{\Delta\rho} = \frac{\sum_{i=1}^{N}\rho(i)}{N} - \rho_{\text{basement}} \tag{3}$$

where $\frac{\sum_{i=1}^{N}\rho(i)}{N}$: is the average density, and

ρ_{basement}: is the density of basement rock.

In the proposed method, the Infinite Horizontal Slab Equation (IHSE) is used to calculate the gravity effects at the earth's surface (or any reference datum) for each subsurface rock' formation that, covering the basement rocks for any sedimentary basin area, and using the IHSE ability, efficiency in the estimating, tracing the formations' thicknesses (or depth), relative to the underlying basement rocks. By using the prior information of control point (or borehole), through profile (s) line (s) of Bouguer gravity anomaly which represents the vertical cross-section (s) for the area of study.

Simply the idea of the new method is based on the assumption that: the sedimentary rock' formations covering the basement rocks are the formations deposited individually in form of layers, or a group of HIS, of different densities distributions is being stacked in columns over the basement's rocks (**Figure 1**). And hence for any point (1, 2, 3, and 4) at the earth's surface (or datum), the total gravity effect is the summation of all gravity effects (at point 4) contributed by each individual slab (1, 2, and 3) along the vertical axis of that point at the earth's surface, were using the average vertical density-contrast ($\overline{\Delta\rho}$), between each formation's slab individually with basement rocks at points 1, 2, 3, and at point 4. Here is the new keen point of view that is: the average vertical densities'-contrasts between the stacked vertically of the IHSs and basement rocks are used in inversion calculations instead of densities'-contrasts between IHSs and their surrounding rock materials, through the

gB Curve of gravity effect along vertical axis

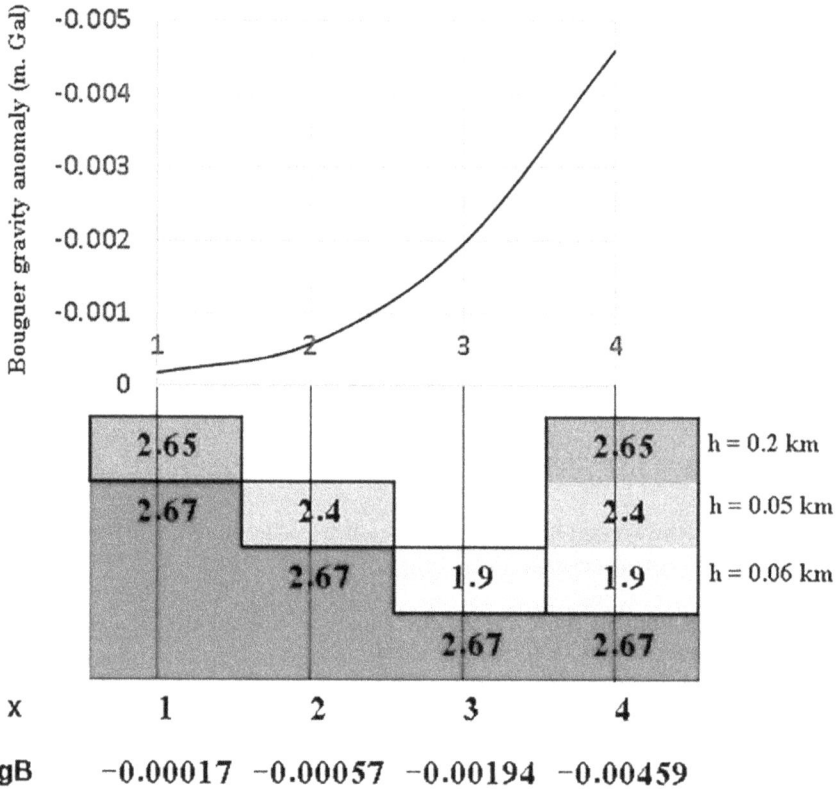

Figure 1.
Bouguer gravity anomalies comparable to all possible rock' formations overlying the basement rocks.

importance for this concept, it being became possible for some extent to separate of Bouguer gravity anomaly to its components that representing probably of all possible rock' formations overlying the basement rocks, according to the prior known information about those rock' formations (thicknesses, depths, and densities), either from subsurface (borehole, etc.) or the surface geology (field, etc.).

2.2 Building two models for formations densities' distributions

To achieve the objective of the newly proposed method, the parameters of the rock' formations of depositions covering the basement rocks for study areas such as their thicknesses, depths, and their densities, should be prior known at least, in one controlling points (borehole data, geophysical data, etc.), and thus such parameters (formations' thicknesses, and densities), can be probably estimated and traced through the Bouguer gravity map's profile (s) from the known point to the other unknown points of the area of study. Hence, a two models for formations density distributions building to prove that, the heterogeneities or homogeneities of formations density distributions do not affect the resultants of depth calculation from gravity effects of the IHSs as follows.

2.2.1 Heterogeneity formations density distributions (model 1)

As shown in **Figure 2**, the proposed model is consisting of a number (N) of deposited layers or formations, deposited according to Walther's Law of deposition of the heterogeneous Juxtaposition of depositions.

For simplification, assuming the model is consisting of five formations (N = 5), and densities (gm/cm^3) from top to bottom are (ρ(N), ρ(N-1), ρ(N-2), ρ(N-3), and ρ(N-4)) with thicknesses (m) are (h1, h2, h3, h4, and h5), respectively. Therefore, the average vertical densities from the top will be as follows:

$$\rho v1(N) = \rho(N)/(N-4) \tag{4}$$

$$\rho v1(N-1) = \rho(N) + \rho(N-1)/(N-3) \tag{5}$$

$$\rho v1(N-2) = \rho(N) + \rho(N-1) + \rho(N-2)/(N-2) \tag{6}$$

$$\rho v1(N-3) = \rho(N) + \rho(N-1) + \rho(N-2) + \rho(N-3)/(N-1) \tag{7}$$

$$\rho v1(N-4) = \rho(N) + \rho(N-1) + \rho(N-2) + \rho(N-3) + \rho(N-4)/(N) \tag{8}$$

Therefore, the average vertical densities for the above modeling is written in form of a row matrix (for the Matlab code purpose) as follows:

$$\overline{\rho v1} = [\rho v1(N) \ \rho v1(N-1) \ \rho v1(N-2) \ \rho v1(N-3) \ \rho v1(N-4)] \tag{9}$$

then the average vertical densities-contrasts are:

$$\overline{\Delta \rho v1}(i) = \overline{\rho v1}(i) - \rho_{\text{basement}} \tag{10}$$

And the gravity effect for model 1, is given as follows:

$$gB_{M1(:,i)} = 2\pi G \overline{\Delta \rho v1}(i)z(i) \tag{11}$$

so that the depths can obtained by the following Eq. (12):

$$Z_{M1(:,i)} = \text{abs}(gB_{M1(i)}/2\pi G \overline{\Delta \rho v1} \tag{12}$$

where $gB_{M1(:,i)}$ are the gravity effects of all points x(i) i.e. all vertical points (i = 1,2,3,4,5), and $Z_{M1(:,i)}$ are the inverted depths at the same vertical points, and also the thicknesses $h_{M1(:,i)}$ have obtained from the following equation:

$$h_{M1(:,)} = \sum_{i=1}^{N} Z_{m1}(i) \tag{13}$$

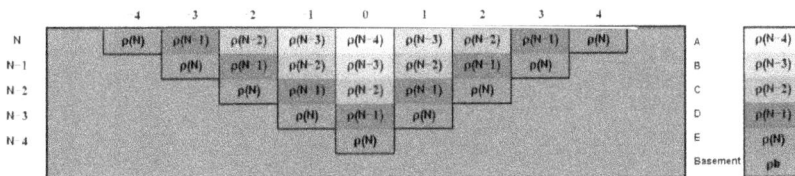

Heterogeneity densities distributions (model-1).

Figure 2.
The model consists of five formations (N = 5), and densities are heterogeneously distributed.

2.2.2 Homogeneity formations density distributions (model 2)

As shown in **Figure 3**, the proposed model is consisting of a number (N) of deposited layers or formations, deposited according to Steno's Law of superposition or Depositional History, Principle of homogeneous Juxtaposition of depositions.

For simplification, assuming the model is consisting of five formations (N = 5), and densities (gm/cm^3) from top to bottom are ($\rho(N)$, $\rho(N-1)$, $\rho(N-2)$, $\rho(N-3)$, and $\rho(N-4)$) with thicknesses (m) are (h1, h2, h3, h4, and h5), respectively. Therefore, the average vertical densities from the top will be as follows:

$$\rho v2(N) = \rho(N)/(N - 4) \tag{14}$$

$$\rho v2(N - 1) = \rho(N) + \rho(N - 1)/(N - 3) \tag{15}$$

$$\rho v2(N - 2) = \rho(N) + \rho(N - 1) + \rho(N - 2)/(N - 2) \tag{16}$$

$$\rho v2(N - 3) = \rho(N) + \rho(N - 1) + \rho(N - 2) + \rho(N - 3)/(N - 1) \tag{17}$$

$$\rho v2(N - 4) = \rho(N) + \rho(N - 1) + \rho(N - 2) + \rho(N - 3) + \rho(N - 4)/(N) \tag{18}$$

Therefore, the average vertical densities for the above modeling is written in form of a row matrix (for the Matlab code purpose) as follows:

$$\overline{\rho v2} = [\rho v2(N)\ \rho v2(N - 1)\ \rho v2(N - 2)\ \rho v2(N - 3)\ \rho v2(N - 4)] \tag{19}$$

then the average vertical densities-contrasts are:

$$\overline{\Delta \rho v2}(i) = \overline{\rho v2}(i) - \rho_{basement} \tag{20}$$

And the gravity effect for model 1, is given as follows:

$$gB_{M2(:,i)} = 2\pi G \overline{\Delta \rho v2}(i) z(i) \tag{21}$$

so that the depths can obtained by the following Eq. (22):

$$Z_{M2(:,i)} = abs\left(gB_{M2(i)}/2\pi G \overline{\Delta \rho v2}\right. \tag{22}$$

where $gB_{M2(:,i)}$ are the gravity effects of all points x(i) i.e. all vertical points (i = 1,2,3,4,5), and $Z_{M2(:,i)}$ are the inverted depths at the same vertical points, and also the thicknesses $h_{M2(:,i)}$ have obtained from the following equation:

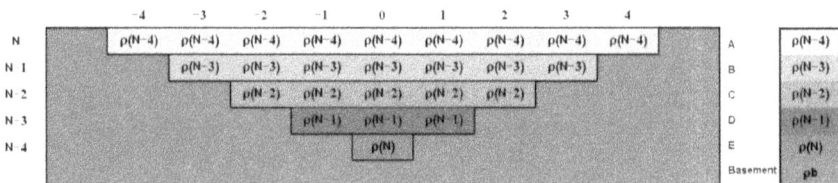

Homogeny densities distributions (model 2).

Figure 3.
The model consists of five formations (N = 5), and densities are homogeneously distributed.

$$h_{M2(:,)} = \sum_{i-1}^{N} Z_{m2}(i) \qquad (23)$$

The goal of geophysical inversion (or interpretation) is to produce models whose response matches observations with noise levels [14]. It is known that the gravity measuring tools are very sensitive only to lateral changes in the causative source, therefore there are several models, that give solutions for the observed profile (ambiguities problem). Even with, this problem the gravity anomalies often are modeled by simple geometrical shapes, (or arbitrary shapes, Talwani et al. 1960). As in all geophysical inversions, there will be ambiguities, notably between density and layer depth, and many of these were pointed out, by [15, 16]. Geophysical inversion by iterative modeling involves fitting observations by adjusting model parameters. Both seismic and potential-field model responses can be influenced by the adjustment of the parameters of rock properties [14].

The new technique in the present research is based on two synthetic models and being built first, consistent, and constrained with real data of known controlling points (or borehole), then applying the algorithm of the solved equations to determine the formations' thicknesses, the basement rocks depth, and tracing them relatively to the formations' thicknesses and depth of basement rocks at the point of a prior known real data, through the profile line of Bouguer anomaly map's covering the area of sedimentary basin.

2.3 Material

The proposed new 2D semi-inversion technique is carried out for any sedimentary basin area, by using a proper corrected Bouguer gravity anomaly map, with a prior known controlling point (s) of the formation's densities and thicknesses (borehole), in addition, using an Excel, Surfer-15 software, and Matlab software for applications the written program for the proposed technique.

2.3.1 Bouguer gravity map

A digitizing process is carried out, for Bouguer's gravity anomaly map that covers the investigated area, processing, and re-contouring with proper contour-intervals Then re-mapping with the location of the prior known controlling point (s) or borehole (s) locations, by using the Surfer-15 software to manipulate and dealing with data easily through the Excel and Matlab software. Thus, then a profile is taken along the map, in digitized form (file with two coordinates (x, gB)) that is later used for algorithm code application in Matlab.

2.3.2 Calculation gravity effects theoretically, with heterogeneous test model 1

The previously, the hypothetical depositional basin model-1 (**Figure 2**) consists of five formations layered slab deposited according to Walther's Law of deposition. Therefore, the formations are filling-basin in five-rows (N = 5), and nine- columns (juxtaposing vertical columns). The formations' thicknesses, depths, and densities are given, as seen in **Table 1**. Where $\Delta\rho v1(i)$, represents here; the average vertical density-contrast for formations, stacked in nine columns of rows numbers N-4, N-3, N-2, N-1, and N respectively, and symmetrically repeated around the maximum formation's thicknesses (central basin where N = 4). By using the equations from (8) to (13) using the Matlab code, the summation values of the vertical effects for stacked slab' columns, at each point at $x(i)$-coordinates ($x(i) = -4, -3, -2, -1, 0, 1,$

Formation	Row No.	Z (km)	h (km)	P (gm/cm³)	V.Av1.ρv1(i) (gm/cm³)	Δρv1(i) (gm/cm³)
A	N - 4	0.5	0.5	1.90	2.7500	0.0800
B	N - 3	1.5	1.0	2.35	2.6500	−0.0200
C	N - 2	1.8	0.3	2.45	2.5833	−0.0867
D	N -1	3.5	1.7	2.55	2.5250	−0.1450
E	N	4.0	0.5	2.75	2.4000	−0.2700
Basement		4.5		2.67		

Table 1.
Data for hypothetical theoretical horizontal slab model (1) of heterogeneous densities distribution.

2, 3, and 4.), are calculated; as well as the formation's thicknesses and depths, are obtained and summarized by following, **Table 2**, where, the formation depths' calculated are: 0.5, 1.5, 1.8, 3.5, and 4.0 km, are corresponding to the thicknesses (h (i)) of each formation sediments in the filling-basin, densities (1.900, 2.350, 2.450, 550, and 2.75 gm/cm³), and the calculated gravity effect curve of the hypothetical sedimentary basin, representing model-1 is seen (**Figure 4**). The depth of the basement is assumed as 4.5 km, and its density is 2.670 gm/cm³.

2.3.3 Calculation gravity effects theoretically, with homogenous test model 2

The previously, hypothetical depositional basin model-2 (**Figure 3**), consists of the same as the previous five formations layered slab deposited according to Walther's and Steno's superposition or geohistory concepts. Therefore, the formations are filling-basin in five rows (N = 5), and nine-columns (juxtaposing vertical columns). The formations' thicknesses, depths, and densities are given, as seen in **Table 3**. Where Δρv2(i), represents here; the average vertical density-contrast for formations, stacked in nine columns of rows numbers N-4, N-3, N-2, N-1, and N respectively, and symmetrically repeated around the maximum formation's thicknesses (central basin where N = 4). By using the equations from (14) to (23) using the Matlab code, the summed values of the vertical effects for stacked slab' columns, at each point at x(i)-coordinates (x(i) = −4, −3, −2, −1, 0, 1, 2, 3, and 4.), are calculated; as well as the formation's thicknesses and depths, are obtained and summarized by following, **Table 4**, where the formations depths' calculated are: 0.5, 1.5, 1.8, 3.5, and 4.0 km, are corresponding to the thicknesses (h (i)) of each formation sediments in the filling-basin, densities (1.900, 2.350, 2.450, 550, and 2.75 gm/cm³), and the calculated gravity effect curve of the hypothetical sedimentary basin, representing model-2 is seen (**Figure 5**). The depth of the basement is assumed as 4.5 km, and its density is 2.670 gm/cm³.

3. Implement the method in cases of real data

3.1 Abu Roash dome area, West Cairo, Egypt (case 1)

The famous Abu Roash Area located between Latitudes 29° 58′ and 30° 03′ N, and longitudes 30° 59′ 10″ and 31° 05′ 19″ E. The Abu Roash district is located 10 km to the southwest of Cairo and is geologically significant because of its surface exposure of Upper Cretaceous rocks [17]. Its name (Abu Roash) is derived from the neighboring village of Abu Roash. The Abu Roash Dome Area constitutes a complex

X (km)	-4	-3	-2	-1	0	1	2	3	4
Av.ρv1(i) (gm/cm3)	2.7500	2.5500	2.5167	2.5250	2.4000	2.5250	2.5167	2.5500	2.7500
Δρv1(i) (gm/cm3)	0.0800	-0.1200	-0.1533	-0.1450	-0.2700	-0.1450	-0.1533	-0.1200	0.0800
gB_M1 (m. Gal)	0.001676	-0.00754	-0.01157	-0.02127	-0.04526	-0.02127	-0.01157	-0.00754	0.001676
h_cal1 (km)	0.5	1.5	1.8	3.5	4	3.5	1.8	1.5	0.5

z_cal1 is calculated average depth of basement $((0.5 + 1.5 + 1.8 + 3.5 + 4)/5) \times 2 = 4.5200$ km.

Table 2.
Theoretical calculation for infinite horizontal slab model (1) for basin filling of five-sedimentary formations overlying basement rocks.

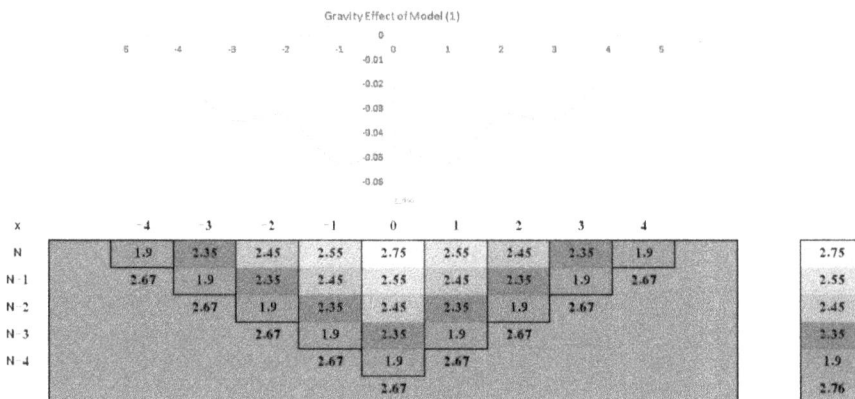

Figure 4.
The calculated gravity effect curve of the hypothetical sedimentary basin, representing model-1.

Formation	Row No.	Z (km)	h (km)	P (gm/cm^3)	V.Av2.ρv2(i) (gm/cm^3)	$\Delta\rho$v2(i) (gm/cm^3)
A	N - 4	0.5	0.5	1.90	2.7500	0.0800
B	N - 3	1.5	1.0	2.35	2.5500	−0.1200
C	N - 2	1.8	0.3	2.45	2.5167	−0.1533
D	N -1	3.5	1.7	2.55	2.5250	−0.1450
E	N	4.0	0.5	2.75	2.4000	−0.2700
Basement		4.5		2.67		

Table 3.
Data for hypothetical theoretical horizontal slab model (2) of homogenous densities distribution.

Cretaceous sedimentary succession with outstanding tectonic features, as shown in **Figure 6**, modified after [18].

3.1.1 Geological setting

The Abu-Roash Dome Area was formed as a result of its location crossing by the western end of the Syrian-arc folds of which extends from northern Egypt to Syria (Laramide orogeny took place in Upper Cretaceous—Lower Tertiary), where the Upper Cretaceous rock formations in the northwestern desert of Egypt had undergone several different tectonic regimes.

The interest in basement depth estimation for the Abu Roash Dome Area was made by several authors' and researchers' gravitational potential studies, such as [19–21]. It is worthily mentioning that, the calculated basement depths calculated by the aforementioned authors' methods, where the depths were estimated from the used modeled body center of a sphere, an infinite long horizontal cylinder, or a semi-infinite vertical cylinder, while in the present method the basement depths are estimated from the top of an infinite horizontal slab.

3.1.2 Procedures and results

The available Bouguer gravity anomaly map (GPC, 1984), covering the Abu Roash Dome Area, was digitized and re-contouring with a proper equal contour

X (km)	-4	-3	-2	-1	0	1	2	3	4
Av.ρv2(i) (gm/cm3)	2.7500	2.5500	2.5167	2.5250	2.4000	2.5250	2.5167	2.5500	2.7500
Δρv2(i) (gm/cm3)	0.0800	-0.1200	-0.1533	-0.1450	-0.2700	-0.1450	-0.1533	-0.1200	0.0800
gB_M2 (m. Gal)	0.00167*	-0.00754	-0.01157	-0.02127	-0.04526	-0.02127	-0.01157	-0.00754	0.001676
h_cal2 (km)	0.5	1.5	1.8	3.5	4	3.5	1.8	1.5	0.5

z_cal2 is calculated average depth of basement ((0.5 + 1.5 + 1.8 + 3.5 + 4)/5) × 2 = 4.5200 km.

Table 4.
Theoretical calculation for infinite horizontal slab model (2) for basin filling of five- sedimentary formations overlying basement rocks.

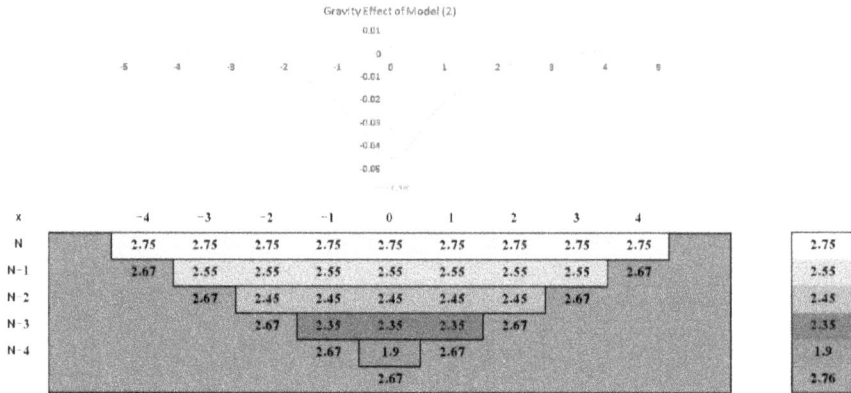

Figure 5.
The calculated gravity effect curve of the hypothetical sedimentary basin, representing model-2.

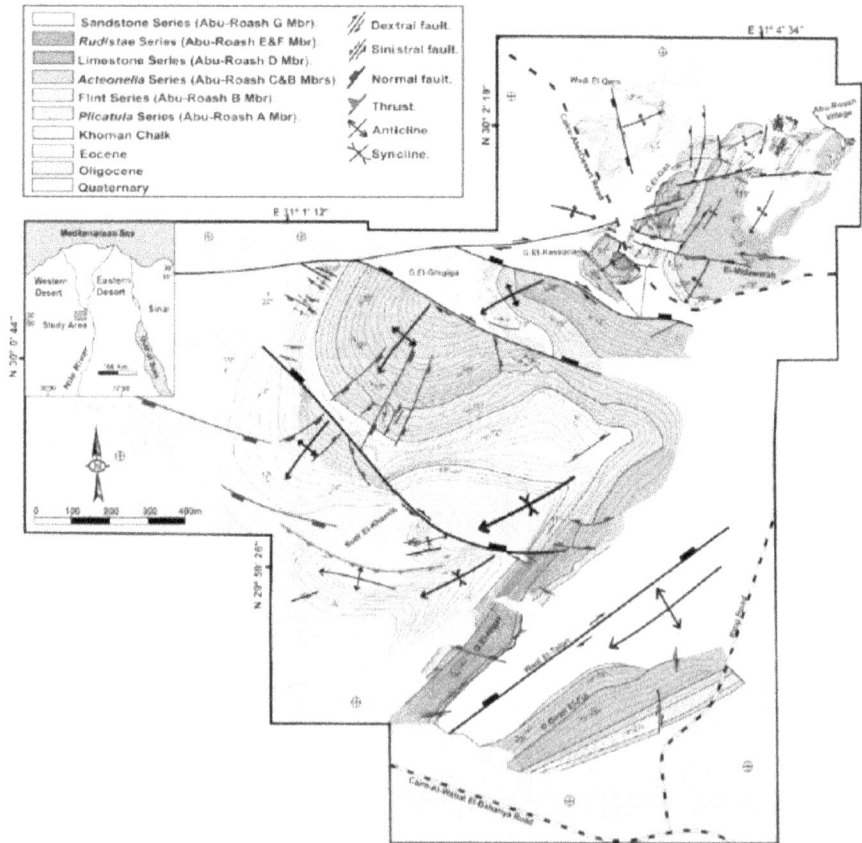

Figure 6.
The location geological setting map of Abu Roash dome area, West Cairo, Egypt.

interval of 2 m. Gal (**Figure 7a**). The Abu Roash-1well, after [22], was used for the interesting formations' depths and corresponding densities were summed in **Table 5** and were used for building the two hypothetical models 1 and 2 for the Abu Roash Dome Area, with heterogeneous and homogeneous formations' densities

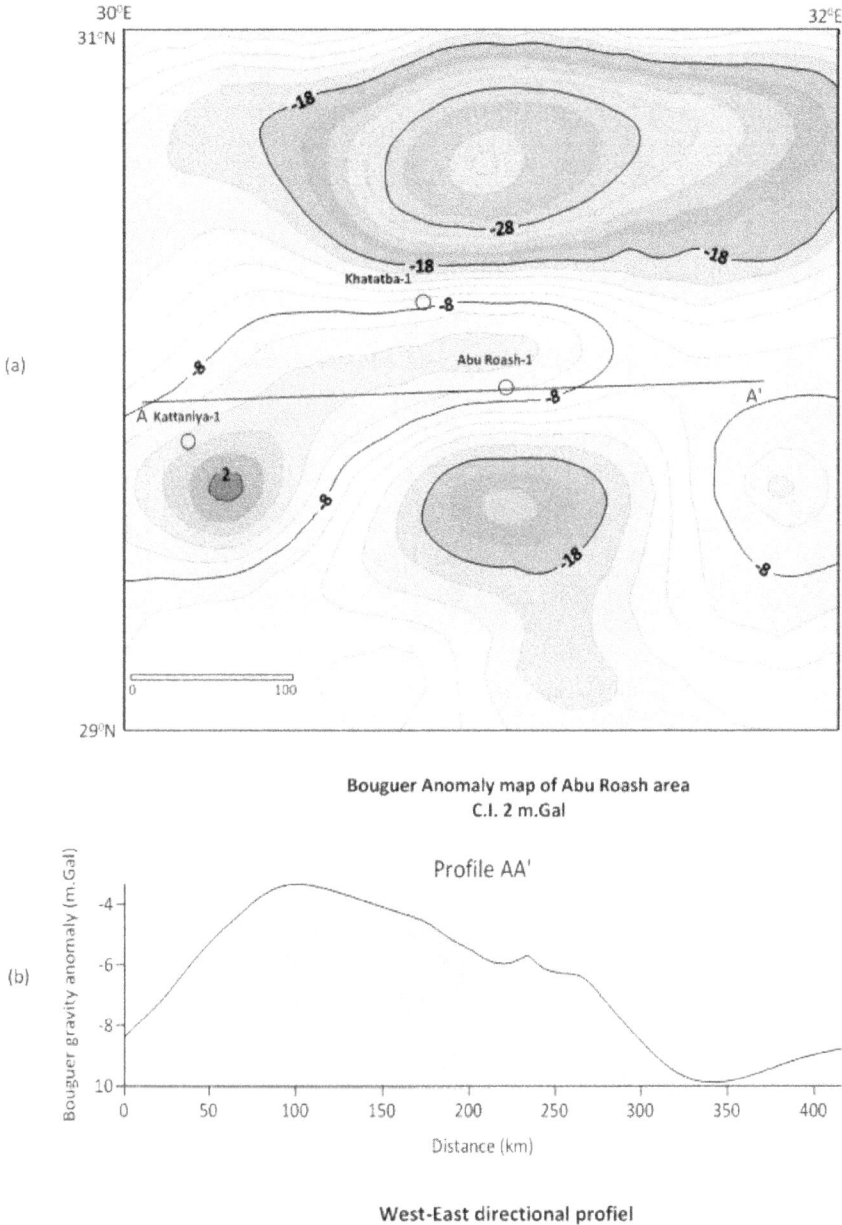

Figure 7.
The bouguer gravity anomaly map of Abu Roash dome area (modified after GPC, 1984).

distributions, as respectively, as shown in **Figures 8** and **9**. For the two models, the theoretical calculations were carried out for thicknesses, averages' vertical densities the averages' vertical densities-contrasts, gravity effect, and calculated thicknesses corresponding to each of the selected five formations, that consists of the Abu Roash Dome area, and were summarized in **Tables 6** and 7.

A digitizing profile along line AA' was carried out for Bouguer gravity anomaly map for Abu Roash Dome Area (**Figure 7b**), with equal intervals 2.09 km., then saved as Excel's file (AbuRoash_aa_slab.xlsx), of two coordinates (x, gB). This file

Formation	Depth (m)	Thickness (m)	Density* (gm/cm³)
Pleistocene	0–161	0.069	1.980
Cenomanian	161–607	446	2.480
Lower Cretaceous	607–759	152	2.610
Jurassic	759–1566	807	2.430
Paleozoic	1566–902	336	2.380
Basement	1.902		2.670

*Densities calculated for lithologic compositions of each formation.

Table 5.
Abu Roash-1 well data (modified after El-Malky, 1985), where the elevation = 92 m and total depth = 1918 m.

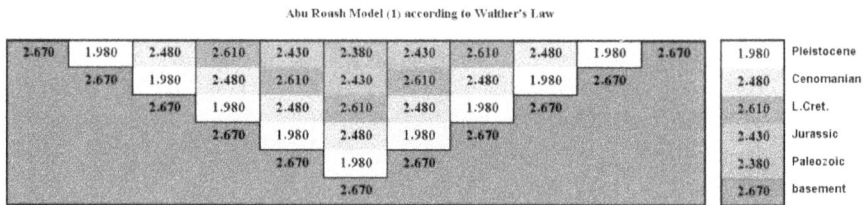

Figure 8.
The hypothetical model 1 for the Abu Roash dome area, with heterogeneous formations' densities distributions.

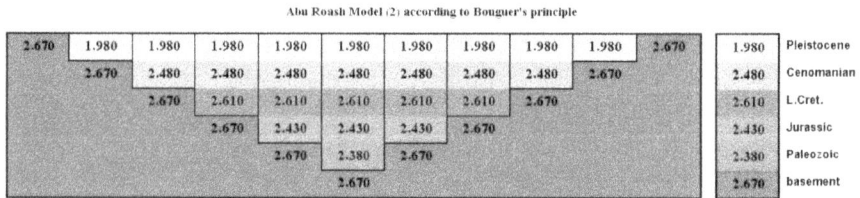

Figure 9.
The hypothetical model 2 for the Abu Roash dome area, with homogeneous formations' densities distributions.

Formation	z (km)	h (km)	ρ (gm/cm³)	ρv1 (gm/cm³)	Δρv1 (gm/cm³)	gB_M1 (m. Gal)	h_M1 (km)
Pleistocene	0.092	0.069	1.980	1.9800	−0.6900	−0.0020	0.0690
Cenomanian	0.161	0.446	2.480	2.2300	−0.4400	−0.0082	0.4460
Lower Cretaceous	0.607	0.152	2.610	2.3567	−0.3133	−0.0020	0.1520
Jurassic	0.759	0.807	2.430	2.3750	−0.2950	−0.0100	0.8070
Paleozoic	1.566	0.336	2.380	2.3760	−0.2940	−0.0041	0.3360
Basement	1.902		2.670				

Table 6.
Abu Roash dome area data and theoretical calculation parameters for model (1).

later will be used for data calculating, tracing the formations' thicknesses, and depth's basement rocks along the profile line AA', by applying the proposed algorithm with Matlab's codes. In the final step, it found that the calculations for the two models along the profile line AA' are given the same results, as expected since the

Formation	z (km)	h (km)	ρ (gm/cm^3)	Pv2 (gm/cm^3)	$\Delta\rho$v2 (gm/cm^3)	gB_M2 (m. Gal)	h_M2 (km)
Pleistocene	0.092	0.069	1.980	1.9800	−0.6900	−0.0020	0.0690
Cenomanian	0.161	0.446	2.480	2.2300	−0.4400	−0.0082	0.4460
Lower Cretaceous	0.607	0.152	2.610	2.3567	−0.3133	−0.0020	0.1520
Jurassic	0.759	0.807	2.430	2.3750	−0.2950	−0.0100	0.8070
Paleozoic	1.566	0.336	2.380	2.3760	−0.2940	−0.0041	0.3360
Basement	1.902		2.670				

Table 7.
Abu Roash dome area data and theoretical calculation parameters for model (2).

x	gB	$h(1)$	$h(2)$	$h(3)$	$h(4)$	$h(5)$	H
0	−8.38956	0.606749	0.386912	0.275528	0.259407	0.258528	1.787124
2.083045	−8.26929	0.598051	0.381366	0.271578	0.255688	0.254822	1.761504
4.166091	−8.15221	0.589583	0.375966	0.267733	0.252068	0.251214	1.736565
6.249136	−8.03704	0.581253	0.370654	0.263951	0.248507	0.247664	1.71203
8.332181	−7.92261	0.572978	0.365377	0.260193	0.244969	0.244138	1.687655
10.41523	−7.80992	0.564828	0.36018	0.256492	0.241484	0.240666	1.66365
12.49827	−7.6972	0.556676	0.354982	0.25279	0.237999	0.237192	1.63964
14.58132	−7.58206	0.548348	0.349671	0.249008	0.234439	0.233644	1.615111
16.66436	−7.46689	0.540019	0.34436	0.245226	0.230878	0.230095	1.590578
18.74741	−7.34443	0.531163	0.338713	0.241204	0.227091	0.226322	1.564493
20.83045	−7.21859	0.522062	0.332909	0.237072	0.2232	0.222444	1.537686
22.9135	−7.0877	0.512596	0.326873	0.232773	0.219153	0.21841	1.509806
–	–	–	–	–	–	–	–
414.526	−8.8006	0.636476	0.405869	0.289028	0.272117	0.271194	1.874683
416.6091	−8.77901	0.634914	0.404873	0.288318	0.271449	0.270529	1.870082
416.6091	−8.77901	0.634914	0.404873	0.288318	0.271449	0.270529	1.870082
		0.473742	0.302097	0.215129	0.202542	0.201855	
	1.387877						1.395366

Table 8.
The thicknesses of formation. Along profile AA' in kilometers (H – Σh(i). i = 1, 2, 3, 4, and 5), Abu Roash dome area.

calculated average vertical density-contrasts are the same for the two models. The results for the two models of Abu Roash Dome Area are summarized in **Tables 8** and **9** representing formations thicknesses, and depths, respectively, and represented graphically as shown in (**Figures 10** and **11**).

3.1.3 Interpretation of data results

From **Table 8**, it was found that the range of formations thicknesses' (minimum to maximum) varying along the profile direction AA' (**Figure 10**), as follows:

x	gB	z(1)	z(2)	z(3)	z(4)	z(5)	Z
0	−8.38956	1.787124	1.787124	1.787124	1.787124	1.787124	1.787124
2.083045	−8.26929	1.761504	1.761504	1.761504	1.761504	1.761504	1.761504
4.166091	−8.15221	1.736565	1.736565	1.736565	1.736565	1.736565	1.736565
6.249136	−8.03704	1.71203	1.71203	1.71203	1.71203	1.71203	1.71203
8.332181	−7.92261	1.687655	1.687655	1.687655	1.687655	1.687655	1.687655
10.41523	−7.80992	1.66365	1.66365	1.66365	1.66365	1.66365	1.66365
12.49827	−7.6972	1.63964	1.63964	1.63964	1.63964	1.63964	1.63964
14.58132	−7.58206	1.615111	1.615111	1.615111	1.615111	1.615111	1.615111
16.66436	−7.46689	1.590578	1.590578	1.590578	1.590578	1.590578	1.590578
18.74741	−7.34443	1.564493	1.564493	1.564493	1.564493	1.564493	1.564493
20.83045	−7.21859	1.537686	1.537686	1.537686	1.537686	1.537686	1.537686
–	–	–	–	–	–	–	–
416.6091	−8.77901	1.870082	1.870082	1.870082	1.870082	1.870082	1.870082
416.6091	−8.77901	1.870082	1.870082	1.870082	1.870082	1.870082	1.870082
		1.388458	1.388458	1.388458	1.388458	1.388458	
	1.387877						1.388458

Table 9.
The depths of formations thickness 'along profile AA' in biometers (Z = Σz(i),i = 1, 2, 3, 4 and 5), Abu Roash dome area.

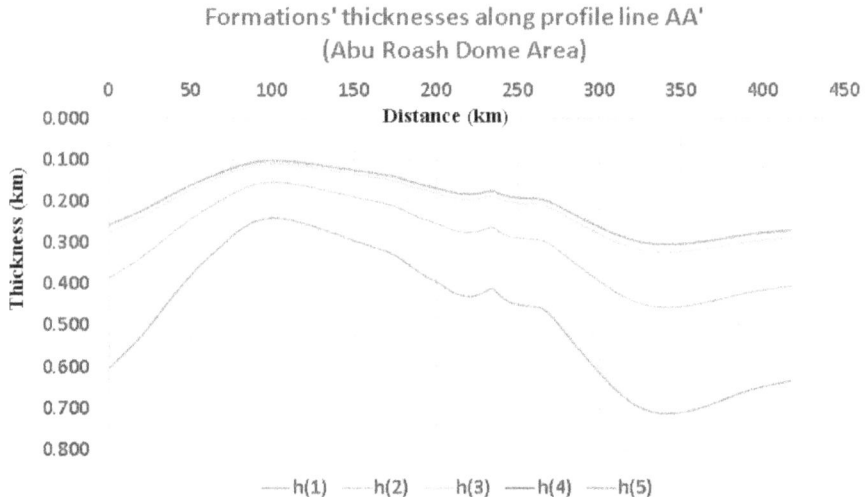

Figure 10.
*The resulted inversion formation' thicknesses along profile AA' of Bouguer map (**Figure 7**).*

- The Pleistocene formation thicknesses range (0.24145–0.71413 km).

- The Cenomanian formation thicknesses range (0.15397–0.45539 km).

- The Lower Cretaceous formation thicknesses range (10964–0.32429 km).

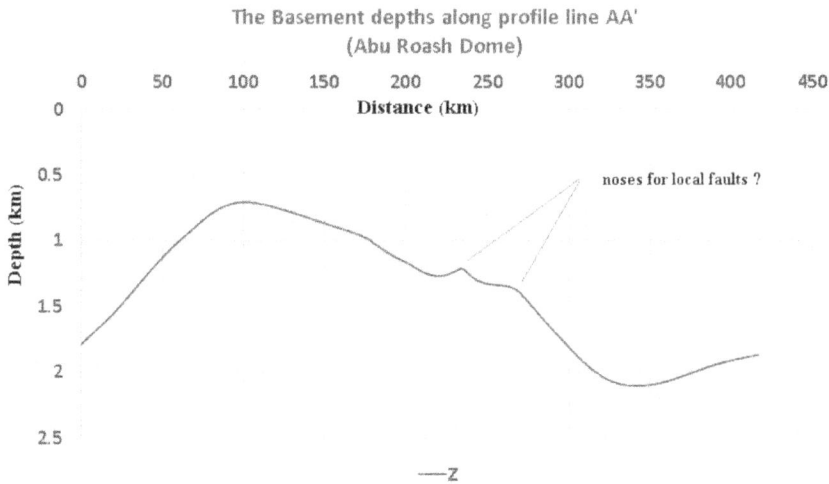

Figure 11.
*The resulted inversion basement rock along profile AA' of Bouguer map (**Figure 7**).*

- The Jurassic formation thicknesses range (0.10323–0.30532 km).

- The Paleozoic formation thicknesses range (0.10288–0.30428 km).

From **Table 7**, the basement depth along the profile line AA' (**Figure 11**), was determined as follows:

The maximum value of average depth (last column in **Table 9**), equal to 2.1034 km, corresponds to the Bouguer anomaly value of about −9.8743 m. Gal, and the minimum value of the last column, equal to 0.7116 km, corresponds to the Bouguer anomaly value of about −3.3385 m. Gal. Therefore, the average basement depth value is 1.40728 km, corresponds to the average Bouguer anomaly −6.6064, this is comparable with depth 1.916 km corresponds to Bouguer anomaly −5.5 m. Gal according to [23]. The calculated basement depths along profile line AA', showed more or fewer values than actual drilled depth (1.902 km), which may be attributed to the lithologic change in the basement rocks, the above overlying sediment thicknesses, and the local faults are indicated as noses on the depths' curve (**Figure 11**).

The Abu Roash Dome depth of value about 2.1 km is obtained by proposed method, that was to some extent agrees with the results obtained from drilling information (1.9 km; after [23], and a S-Curves method of depth determination (1.91 km; after [24]).

3.2 Humble salt dome in Harries County, Texas USA (case 2)

The gravimetric survey, with its sensitivity to variations in density-contrast among the subsurface structures, has been helpful in discovering and locating salt dome formations common to the Gulf Coast, of the USA.

3.2.1 Geological setting

The salt domes considering interesting as a source producing oils, minerals like Sulfur, Salts, and recently are used as burial locations for waste disposal of nuclear

129

materials. Salt domes are common in the Gulf Coast area of Texas and Louisiana as well as in the Gulf of Mexico. The Gulf of Mexico basin began forming in the late Triassic as an intracontinental extension within the North American plate [25].

Salt was accumulated in the Jurassic period and geologically identified as the Louann Salt (mother source), which is a very thick deposit of salt known as halite composed of sodium chloride but with smaller amounts of sulfate, halides, and borates. The salt was followed by carbonate deposition during the Late Jurassic and Cretaceous, and clastic deposits during the Cenozoic [26]. With the deposition of additional sediments on top of this salt, it was buried to over 20,000 feet (6.096 km) and sometimes as deep as 40,000 feet (12.192 km) in the Deep-Water Gulf of Mexico.

The Humble Salt Dome in Harris County, Texas, USA, is one of the interiors of the Gulf Coastal Plain (**Figure 12**), and it is more than 20,000 feet (6.096 km) in diameter and less than 2000 feet (0.6096 km) below the surface.

The Humble Salt Dome estimation depth has been subjected to studies from several authors such as [21, 27–38]. Also, it is worthily mentioning that the calculated salt dome depths calculated by the aforementioned authors' methods, depths were estimated from being considering the shape modeled salt body's center either of a sphere, an infinite long horizontal cylinder, or a semi-infinite vertical cylinder, while in the present method the salt's depths are related to basement rocks depths' and are being estimated from the top of an infinite horizontal slab.

3.2.2 Procedures and results

The available Bouguer gravity anomaly map of Humble Salt Dome, Harris County, Texas, USA Area (After [27]), is digitized and re-contouring with a proper equal contour interval of 2 m. Gal (**Figure 13a**), where the gravity anomalies range

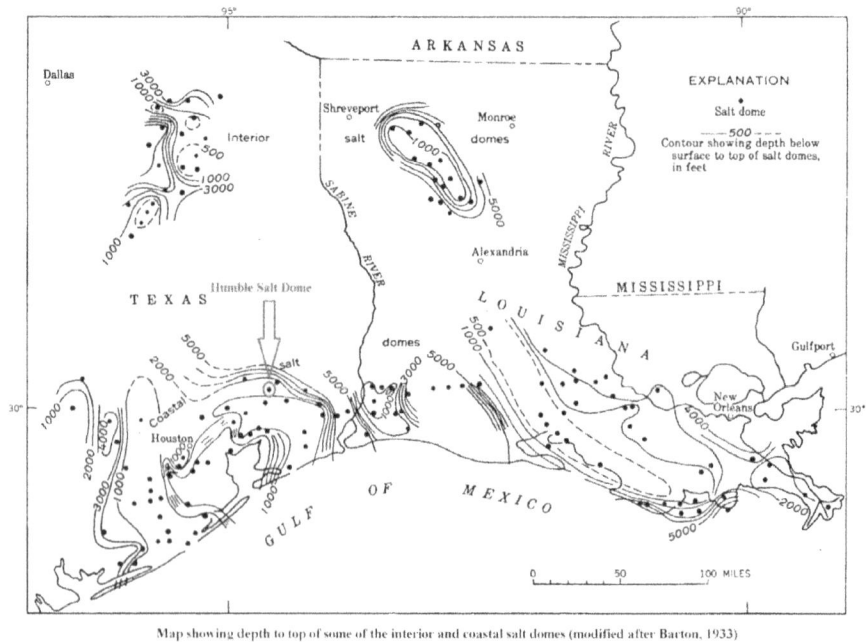

Map showing depth to top of some of the interior and coastal salt domes (modified after Barton, 1933)

Figure 12.
The location of humble salt dome referred in red circle on depth map (contours in feet).

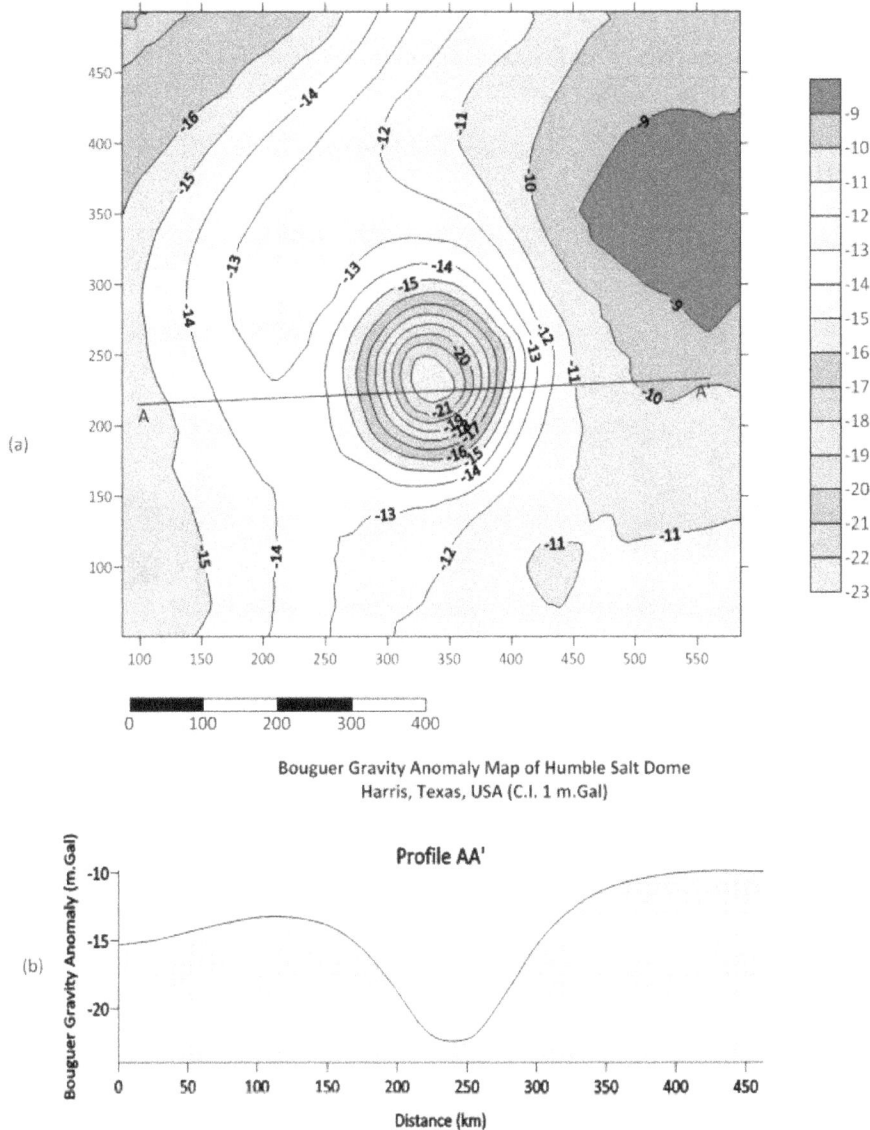

Figure 13.
The bouguer gravity anomaly map of humble salt dome (modified after Nettleton 1976).

between −9 m. Gal at the northeastern part of the map and more slightly of the value −22 m. Gal, at the center of Humble Salt Dome.

The stratigraphic of formations, depths, thicknesses, and densities, as a controlling-point are obtained (after, [39]), were summed in **Table 10** and was used for building the two hypothetical models 1 and 2 for the Humble Salt Dome. model (1) with heterogeneous formations' densities distributions and model (2) with homogeneous formations' densities distributions, as respectively shown (**Figures 14** and **15**). For the two models, the theoretical calculations were carried for thicknesses, averages' vertical densities the averages' vertical densities-contrasts, gravity effect, and calculated thicknesses corresponding to each of the selected five formations, which consists of the Humble Salt Dome, and was summarized in **Tables 10** and **11**.

Formation	Lithology	Depth (m)	Thickness (m)	Density *(gm/cm^3)
A	Clay, Shale, Silt, and Sand	182.88	1371.6	2.510
B	Clay, Shale, and Sand	1554.48	274.32	2.670
C	Limestone and Shale	1828.80	548.64	2.700
D (Salt)	Rock Salt	2377.44	822.96	2.100
Basement	Granitic—Dioritic	3048.00		2.950

Densities are calculated for lithologic compositions of each formation.

Table 10.
Data information for control-point of humble salt dome (modified after, Okocha, 2017).

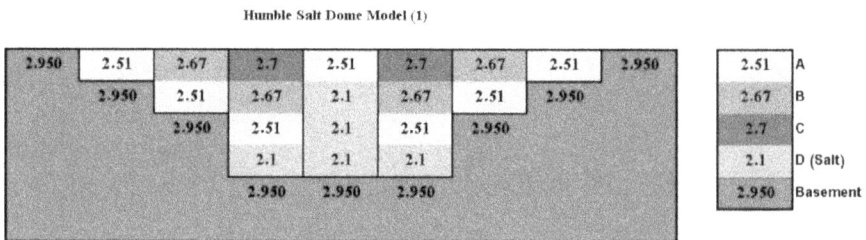

Figure 14.
The hypothetical model 1 for the humble salt dome, with heterogeneous formations' densities distributions.

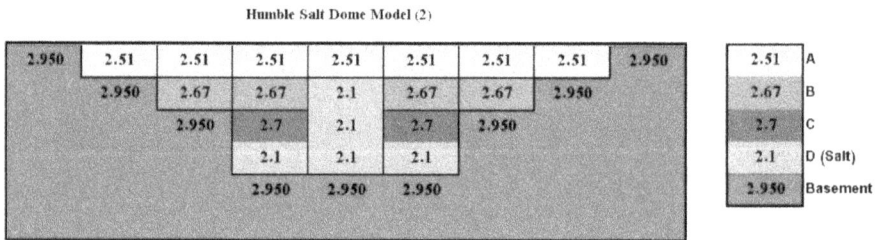

Figure 15.
The hypothetical model 2 for the humble salt dome, with homogeneous formations' densities distributions.

Formation	z (m)	h (m)	ρ (gm/cm^3)	ρv1 (gm/cm^3)	Δρv1 (gm/cm^3)	gB_M1 (m. Gal)	h_M1 (m)
A	182.88	182.880	2.51	2.5100	−0.4400	−0.0253	1.3716
B	1554.48	1371.600	2.67	2.5900	−0.3600	−0.0041	0.27432
C	1828.8	274.320	2.7	2.4950	−0.4550	−0.0105	0.54864
D (Salt)	2377.44	548.640	2.1	2.2025	−0.7475	−0.0258	0.82296
Basement	3078		2.95				

Table 11.
Humble salt dome and theoretical calculation parameters for model (1).

A digitizing profile along line AA' was carried out for Bouguer gravity anomaly map for Humble Salt Dome (Figure 14b), with equal intervals 0.35253 km., where the profile AA' is about 30.5 km in length. Then the digitized values are saved as Excel's file (humble_aa_slab.xlsx), of two coordinates (x, gB), where the file was later used for calculating, tracing the formations' thicknesses, and depth's basement rocks along the profile line AA', by applying the algorithm of the proposed code with Matlab's. In the final step, it found that the calculations for the two models along the profile line AA' are given the same results, as expected since the calculated average vertical density-contrasts is the same for the two models. The results for the two models of Humble Salt Dome are summarized in **Tables 12** and **13** representing formations thicknesses, and depths, respectively, and represented graphically as shown in (**Figures 16** and **17**).

3.2.3 Interpretation of data results

From **Table 12**, it was found that the range of formations thicknesses (minimum to maximum) varying along the profile direction AA' (**Figure 16**), as follows:

Formation	z (m)	h (m)	ρ (gm/cm^3)	Pv2 (gm/cm^3)	$\Delta\rho$v2 (gm/cm^3)	gB_M2 (m. Gal)	h_M2 (m)
A	182.88	182.880	2.51	2.5100	−0.4400	−0.0253	1.3716
B	1554.48	1371.600	2.67	2.5900	−0.3600	−0.0041	0.27432
C	1828.8	274.320	2.7	2.4950	−0.4550	−0.0105	0.54864
D (Salt)	2377.44	548.640	2.1	2.2025	−0.7475	−0.0258	0.82296
Basement	3078		2.95				

Table 12.
Humble salt dome and theoretical calculation parameters for model (2).

x	gB	$h(1)$	$h(2)$	$h(3)$	$h(4)$	H
0	−15.30482343	0.706	0.577	0.730	1.199	3.211
2.311662826	−15.27975324	0.704	0.576	0.728	1.197	3.206
4.623325652	−15.25370612	0.703	0.575	0.727	1.195	3.200
6.934988478	−15.22756542	0.702	0.574	0.726	1.193	3.195
9.246651304	−15.20155964	0.701	0.573	0.725	1.191	3.189
11.55831413	−15.17557184	0.700	0.572	0.723	1.189	3.184
13.86997696	−15.14762043	0.698	0.571	0.722	1.186	3.178
16.18163978	−15.11822282	0.697	0.570	0.721	1.184	3.172
18.49330261	−15.0881731	0.696	0.569	0.719	1.182	3.166
20.80496543	−15.05724459	0.694	0.568	0.718	1.179	3.159
23.11662826	−15.02432933	0.693	0.567	0.716	1.177	3.152
25.42829108	−14.98468037	0.691	0.565	0.714	1.174	3.144
–	–	–	–	–	–	–
460.0209024	−9.891211115	0.456	0.373	0.472	0.775	2.075
462.3325652	−9.893079595	0.456	0.373	0.472	0.775	2.076
462.3325652	−9.893079595	0.456	0.373	0.472	0.775	2.076
		134.925	110.393	139.525	229.219	
	3.039908					3.039908

Table 13.
The thickness of formations along profile AA' in kilometers (H = Σh(i), i = 1, 2, 3, 4, and 5), humble salt dome.

Formations Thickness' along profile AA'
(Humble Salt Dome)

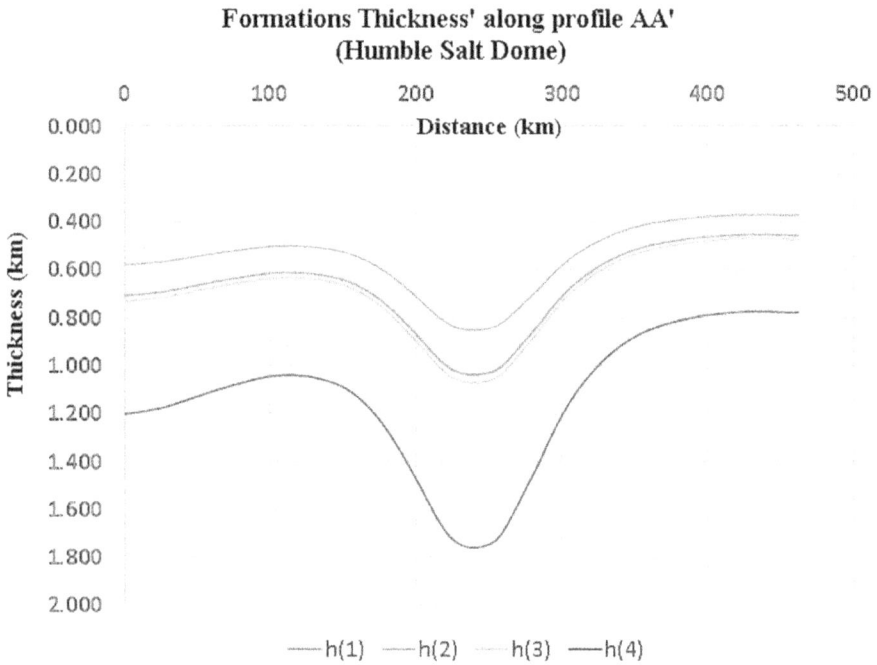

Figure 16.
*The resulted inversion formation' thicknesses along profile AA' of bouguer map (**Figure 13**).*

Basement depths along profile line AA' Humble Salt

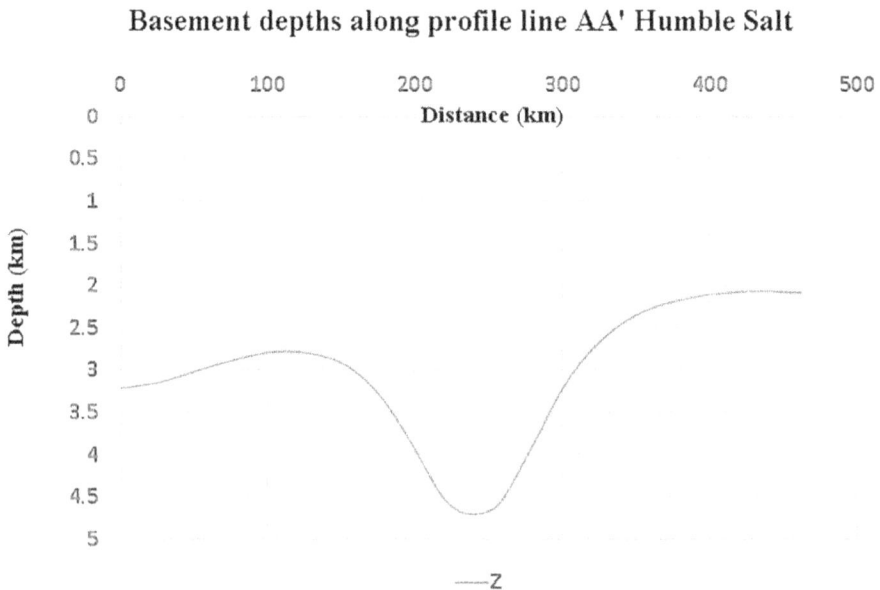

Figure 17.
*The resulted inversion basement rock along profile AA' of bouguer map (**Figure 13**).*

- The A-formation thicknesses range (0.453–1.035 km).

- The B-formation thicknesses range (0.371–0.847 km).

- The C-formation thicknesses range (0.469–1.070 km).

- The D (Salt)-formation thicknesses range (0.770–1.758 km).

From **Table 13**, the basement depth along the profile line AA' (**Figure 9**), was determined as follows:

The maximum value of average depth (last column in **Table 13**), equal to 4.70952 km, corresponds to the Bouguer anomaly value of about −22.437979 m. Gal (near the center of Salt Dome anomaly), and the minimum value of the last column, equal to 2.06245 km, corresponds to the Bouguer anomaly value of about −9.826328 m. Gal (near the edges of bounded the Salt Dom anomaly). Therefore, the average depth to the center of the Humble Salt Dome is about 3.386 km, corresponds to the Bouguer anomaly of about −16.128 m. Gal.

The Humble Salt Dome depth of value about 4.71 km is obtained by proposed method, that was agrees very well with the results obtained from drilling, seismic information (4.97 km; after [27]), and a simple method of depth determination by using shape factor (4.85 km; after [21]).

4. Discussion and conclusions

The present research represents a new "semi-inversion" method for calculating, and tracing formations' thicknesses and basement rocks depths, for deposition basin, relatively to a prior known control-point (s) or borehole (s), throughout profile (s) line (s) of Bouguer gravity anomaly map that connecting with the controlling-point (s). The present technique is to mimic to some extent tracing formations from borehole data to the seismic cross-section, in the seismic interpretation process. The resulting thicknesses and/or depths for profile (s) line (s) of Bouguer map, covering any being investigated area might be reused again to form grids for any interesting formation concerning the subsurface. Theoretical and field examples reveal the goodness and the efficiency of the method presented. Moreover, the method can be developed and used to help with planning seismic surveys.

4.1 The most important of advantages and disadvantages of proposed method

The new method has several advantages, more than other traditional separation methods. The most important is its capability for separation of Bouguer gravity anomaly above any depositional basin to directly its formations layers, and tracing them from a known point. But in the other methods it being separate only components of regional (basement rock or deeper) and residual (sediments rocks or shallower). On the other hand, side, the method is considered a pioneer theoretically, but still need an effort to develop and optimize of the Matlab Programming code, to be more efficient, saving time, and money in practical application.

Acknowledgements

The author sincerely first thanks in advance to Dr. Karmen Daleta, Author Service Manager, for his kindest invitation, his follow during revision, Dr. Editor-Chief, Editors, Publishing Processor Manager, for helping, and directing during writing this paper. The gratitude extends to Prof. Dr. Khaled Essa for invitation and acceptance for sharing in IntechOpen-Chapter, and for his encouragement. Also, gratitude extended to my family especially my wife, for providing a good environment for carrying for this work.

Appendix

```
%Semi-Inversion_Finite_Slab_Model
% the method is depending on the concepts of Walther's Law of deposition
% and the Steno's Law of superposition of the juxtaposed columns of deposition
clc; % Clear the command window.
close all; % Close all figures (except those of in tool.)
clear; % Erase all existing variables. Or clear it if you want.
workspace; % Make sure the workspace panel is showing.
G = 6.67e-3; % universal gravitational constant (6.67e-3);
pi = (22/7); % circle D/R ratio or solid angle.
format long % for accurate decimal values
% gz in m. Gal//G = 6.67e-11//density contrast in (kg/m.^3)// (x, z, R) in (m)
%=============================================================%
%                          Modeling Parameters
%=============================================================%
N = 5 ;     % the number of stacked H. layers = the number of stacked H. layers
        % (The maximum columns will contain the 5-Formations)
%=============================================================%
%                          Reading Data File
%=============================================================%
data = xlsread('Abu Roach_aa_slab.xlsx');
xc = data(:,1);
% the observed points for digitized profile (km)
gB = data(:,2);
% digitized Bouguer anomaly data (profile) (m. Gal)
%i = 202;
x = 0: N-1;
%=============================================================%
%                          Borehole depths (km)
%=============================================================%
z1 =0.092; % Surface of Earth
z2 =0.161; % Cenomanian Fm.
z3 =0.607; % L. Cret. Fm.
z4 =0.759; % Jurassic Fm.
z5 =1.566; % Paleozoic Fm.
      z6 = 1.902; % T. Depth Fm.
z = [z1, z2, z3, z4 z5] ;
% measuring depths from datum sea level (L.S.(z=0) )
%=============================================================%
%                          Borehole thicknesses (km)
%=============================================================%
h1 = z2-z1;
h2 = z3-z2; % Cenomanian Fm.
h3 = z4-z3; % L. Cret. Fm.
h4 = z5-z4; % Jurassic Fm.
h5 = z6-z5; % Paleozoic Fm.
h= [h1, h2, h3, h4 h5] ; % ok
```

```
%================================================================%
%                    Borehole vertical accumulated thickness
%================================================================%
h_v1 = h1;
h_v2 = h1+h2;
h_v3 = h1+h2+h3;
h_v4 = h1+h2+h3+h4;
h_v5 = h1+h2+h3+h4+h5;
h_v = [h_v1 h_v2 h_v3 h_v4 h_v5];
% measuring depths from datum surface level (L.S.(z=0.092) )
% the depth to the central bottom z2(j)(km)
% h(i) are the thicknesses of formations (km)
%================================================================%
%                    Borehole Densities (gm/cm^3)
%================================================================%
rho(N-4)= 1.980; % Pleistocene Fm. 1 (gm/cm^3)
rho(N-3)= 2.480; % Cenomanian Fm. 2 (gm/cm^3)
rho(N-2)= 2.610; % L.Cret. Fm. 3 (gm/cm^3)
rho(N-1)= 2.430; % Jurassic Fm. 4 (gm/cm^3)
rho(N) = 2.380; % Paleozoic Fm. 5 (gm/cm^3)
rho_basement = 2.67; % basement rock 6 (gm/cm^3)
%================================================================%
rho_v1(N-4)= (rho(N-4))/(N-4);
rho_v1(N-3)= (rho(N-4)+ rho(N-3))/(N-3);
rho_v1(N-2)= (rho(N-4)+ rho(N-3)+ rho(N-2))/(N-2);
rho_v1(N-1)= (rho(N-4)+ rho(N-3)+ rho(N-2)+ rho(N-1))/(N-1);
rho_v1(N) = (rho(N-4)+ rho(N-3)+ rho(N-2)+ rho(N-1)+ rho(N))/(N);
rho_v1 = [rho_v1(N-4) rho_v1(N-3) rho_v1(N-2) rho_v1(N-1) rho_v1(N)];%ok
delta_rho1 = rho_v1 - rho_basement; %ok
%================================================================%
rho_v2(N-4)= (rho(N-4))/(N-4);
rho_v2(N-3)= (rho(N-4)+ rho(N-3))/(N-3);
rho_v2(N-2)= (rho(N-4)+ rho(N-3)+ rho(N-2))/(N-2);
rho_v2(N-1)= (rho(N-4)+ rho(N-3)+ rho(N-2)+ rho(N-1))/(N-1);
rho_v2(N) = (rho(N-4)+ rho(N-3)+ rho(N-2)+ rho(N-1)+ rho(N))/(N);
%============================================%
rho_v2 = [rho_v2(N-4) rho_v2(N-3) rho_v2(N-2) rho_v2(N-1) rho_v2(N)];%ok
delta_rho2 = rho_v2 - rho_basement; %ok
%================================================================%
%                    conditions for calculations
%================================================================%
%                    Theoritical_Salb calculations
%================================================================%
i = zeros();
for i = 1: N;
%================================================================%
%                    Model (1) Historical Concept
%================================================================%
gB_M1(:,i)= 2*pi()*G*delta_rho1(i)*h(i);
% gravity effect of slab_model (1) ok
z_M1(i)= abs(gB_M1(i)/(2*pi()*G*delta_rho1(i)));
% thicknesses of formation (km) ok
h_calM1(:,i) = sum(z_M1(i)); % depth of formation
```

```
delt_gB1 = 2*pi()*G*delta_rho1(i); % rate of anomaly change with thickness
%=========================================================%
%                        Model (2) Bouguer Concept
%=========================================================%
gB_M2(:,i)= 2*pi()*G*delta_rho2(i)*h(i);
% gravity effect of slab_model (2) ok
z_M2(i)= abs(gB_M2(i)/(2*pi()*G*delta_rho2(i)));
% thicknesses of formation (km) ok
h_calM2(:,i) = sum(z_M2(i)); % depth of formation
delt_gB2 = 2*pi()*G*delta_rho2(i); % rate of anomaly change with thickness
%=========================================================%
%                        Profile Calculations
%=========================================================%
% Model (1)
h_v_cal1= (abs((gB/2*pi()*G*delta_rho1))*10);
% thicknesses of formation (km)
z_cal1(:,i) = (sum(h_v_cal1, 2));
% maximum depth (depth to the basement) (km)
%=========================================================%
% Model (2)
h_v_cal2= (abs((gB/2*pi()*G*delta_rho2))*10);
% thicknesses of formation (km)
z_cal2(:,i) = (sum(h_v_cal2, 2));
% maximum depth (depth to the basement) (km)
%=========================================================%
end
%=========================================================%
%             Graphical Representations Model (1)& Model (2)
%=========================================================%
figure(1)
plot(x,gB_M1,'k-')
hold on
grid on
set(gca, 'YDir','reverse')
xlabel('xc -axis of measured Bouguer (km)');
ylabel('gravity anomaly gB_Model(1) in (m.Gals)');
title('Gravity anomaly over horizontal slabs')
%=========================================================%
figure(2)
plot(x,z_M1,'k-')
hold on
grid on
set(gca, 'YDir','reverse')
xlabel('xc -axis of measuered Bouguer (km)');
ylabel(' calculated h_M1 Model (1) depth in (km)');
title('calculated thickness using slab model')
%=========================================================%
figure(3)
plot(x,h_calM1,'k-')
hold on
grid on
set(gca, 'YDir','reverse')
xlabel('xc -axis of measuered Bouguer (km)');
```

```
ylabel(' calculated z_calM1 Model (1) thickness in (km)');
title('calculated depth using slab model')
%===========================================================%
figure(4)
plot(x,gB_M2,'k-')
hold on
grid on
set(gca, 'YDir','reverse')
xlabel('xc -axis of measuered Bouguer (km)');
ylabel('gravity anomaly gB_Model(2) in (m.Gals)');
title('Gravity anomaly over horizontal salbs')
%===========================================================%
figure(5)
plot(x,z_M2,'k-')
hold on
grid on
set(gca, 'YDir','reverse')
xlabel('xc -axis of measuered Bouguer (km)');
ylabel(' calculated h_M1 Model (2) depth in (km)');
title('calculated thickness using slab model')
%===========================================================%
figure(6)
plot(x,h_calM2,'k-')
hold on
grid on
set(gca, 'YDir','reverse')
xlabel('xc -axis of measuered Bouguer (km)');
ylabel(' calculated z_calM2 Model (2) thickness in (km)');
title('calculated depth using slab model')
```

Author details

D.M. Abdelfattah
Former J.V. Geophysicist-Eni's Company in Egypt, Egypt

*Address all correspondence to: moustafa_dahab@yahoo.com

IntechOpen

References

[1] Sharma PV. Environmental and Engineering Geophysics. Cambridge, UK: Cambridge University Press; 1997

[2] Goussev SA et al. Gravity and Magnetics Exploration Lexicon. Calgary, Alberta, Canada: Geophysical Exploration & Development Corporation (GEDCO); 2000

[3] Parker RL. The theory of ideal bodies for gravity interpretation. Geophysical Journal Royal Astronomical Society. 1975;**42**:315-334

[4] Parker RL. Understanding inverse theory. Annual Review of Earth and Planetary Sciences. 1977;**5**:35-64 Available from: www.annualreviews.org/aronline by California Institute of Technology on 10/03/07. For personal use only

[5] A Salem et al. Estimation of depth and shape factor from potential-field data over sources of simple geometry, Conference Paper in SEG Technical Program Expanded Abstracts Society of Exploration Geophysicists United States 2002, DOI: 10.1190/1.1817358. 2002.

[6] Werner S. Interpretation of Magnetic Anomalies at Sheet-like Bodies. Sweden: Sveriges Geologiska Undersok, Series C, Arsbok; 1953

[7] Reid AB et al. Magnetic interpretation in three dimensions using Euler deconvolution. Geophysics. 1990;**55**:80-91

[8] Salem A et al. Generalized magnetic tilt-euler deconvolution. In: SEG Technical Program Expanded Abstracts. Richardson, Texas: OnePetro; 2007. DOI: 10.1190/1.2792530

[9] Fedi M et al. Understanding imaging methods for potential field data. Geophysics. 2012, 77;(1):13. DOI: 10.1190/geo2011-0078.1

[10] Lillie RJ. Whole earth geophysics: An introduction textbook for geologists and geophysicists. In: Library of Congress Cataloging-in-Publication Data. ISBN: 0-13-490517-2. USA: Prentice-Hall, Inc; 1999

[11] Phillips JD et al. National Geophysical Data Grids; Gamma Ray, Gravity, Magnetic, and Topographic Data for the Conterminous. United States: US Geological Survey Digital Data Series DDS-09, 1 CD-ROM; 1993

[12] Nozaki K. The generalized bouguer anomaly. Earth, Planets and Space. 2006;**58**:287-303

[13] Reynolds M. An Introduction to Applied and Environmental Geophysics. Chichester, West Sussex, England: John Wiley & Son Ltd; 1997

[14] Lines LR et al. . In: Fitterman DV, Ellefsen K, editors. Fundamentals of Geophysical Interpretation, Geophysical Monograph Series. Vol. 13. Tulsa, Oklahoma, United States. ISBN: 978-0-931830-56-3 (Series): Society of Exploration Geophysicists; 2004

[15] Skeels D. C. Ambiguity in gravity interpretation. Geophysics. 1947;**12**: 43-56

[16] David Johnson B et al. Some equivalent bodies and ambiguity in magnetic and gravity interpretation. Exploration Geophysics. 1979;**10**(1): 109-110

[17] Abdel Khalek ML et al. Structural history of Abu Roash district, Western Desert, Egypt. Journal of African Earth Sciences. 1989;**9**(3/4):435-443

[18] Shided GA et al. Land use change detection and urban extension at Abu-Roash environs, West Cairo, Egypt. Nessa Publishers (NR). Journal of Geology & Earth Sciences. 2019;**1**(4):13

[19] Abdelrahman EM et al. On the least-squares residual anomaly determination. Geophysics. 1985;**50**(3):473-480. DOI: 10.1190/1.1441925

[20] H El-Araby et al. An iterative approach to depth determination of a buried sphere from vertical magnetic anomalies. Arab Gulf Journal of Scientific Research. 1993;**11**(1):37-46

[21] Khalid S, ESSA. A simple formula for shape and depth determination from residual gravity anomalies. Acta Geophysica. 2007;**55**(2):182-190. DOI: 10.2478/s11600-007-0003-9

[22] El-Malky GM. Structural Analysis of the Northern Western Desert of Egypt [PhD. thesis]. , London S.W.7: University of London, Structure Geology Section, Department of Geology, Royal School of Mines, Imperial College; 1985

[23] Sharaf M. The distribution of sedimentary basins in North Egypt and their relation to basement tectonics. Delta Journal of Science. 1988;**12**(2): 539-560

[24] Khalid S, ESSA. Gravity data interpretation using the S-Curves method. Journal of Geophysics and Engineering. 2007;**4**(2):204-213. DOI: 10.1088/1742-2132/4/2/009

[25] Salvador A. Triassic-Jurassic. In: Salvador A, editor. The Gulf of Mexico Basin. Boulder, Colorado: Geological Society of America; 1991h, pp. 131-180

[26] Kupfer DH. Environment and intrusion of Gulf Coast salt and its probable relationship to plate tectonics. In: Fourth Symposium on Salt, Cleveland, Ohio. Northern Ohio: Geological Society; 1974. pp. 197-213

[27] Nettleton LL. Gravity and Magnetics in oil Prospecting. New York, NY, USA: McGraw-Hili Book Co; 1976

[28] Rao M et al. Two-dimensional interpretation of gravity anomalies over sedimentary basins with an exponential decrease of density-contrast with depth. Journal of Earth System Science. 1999; **108**(2):99-106

[29] Abdelrahman EM et al. Gravity interpretation using correlation factors between successive least-squares residual anomalies. Geophysics. 1989; **54**(12):1614-1621. DOI: 10.1190/ 1.1442629

[30] Abdelrahman EM et al. A least-squares minimization approach to depth determination from moving average gravity anomalies. Geophysics. 1993;**59**: 1779-1784

[31] Abdelrahman EM et al. A numerical approach to depth determination from residual gravity anomaly due to two structures. Pure and Applied Geophysics. 1999;**154**(2):329-341. DOI: 10.1007/s000240050232

[32] Abdelrahman EM et al. Three least-squares minimization approach to depth, shape, and amplitude coefficient determination from gravity data. Geophysics. 2001;**66**:1105-1109

[33] Salem A, Ravat D. A combined analytic signal and Euler method (AN-EUL) for automatic interpretation of magnetic data. Geophysics. 2003;**68**: 1952-1961

[34] Salem A, Elawadi E, Ushijima K. Detection of cavities and tunnels from gravity data using a neural network. Exploration Geophysics. 2004;**32**: 204-208

[35] TLAS M et al. A versatile nonlinear inversion to interpret gravity anomaly caused by a simple geometrical structure. Pure and Applied Geophysics. 2005;**162**:2557-2571

[36] Aghajani H et al. Normalized full gradient of gravity anomaly method and

its application to the mobrun sulfide body Canada. World Applied Sciences Journal. 2009;**6**(3):392-400. ISSN: 1818-4952. IDOSI Publications. 2009

[37] Asfahani J et al. Fair function minimization for direct interpretation of residual gravity anomaly profiles due to spheres and cylinders. Pure Applied Geophysics. 2012;**169**(1):157-165

[38] Mehanee SA. Accurate and efficient regularized inversion approach for the interpretation of isolated gravity anomalies. Pure and Applied Geophysics. 2014;**171**:1897e1937. DOI: 10.1007/s00024-013-0761-z

[39] Okocha, C. Gravitational Study of the Hastings Salt Dome and Associated Faults in Brazoria and Galveston Counties, Texas [M.Sc. Thesis]. Nacogdoches, TX, United States: Stephen F. Austin State University; 2017

Chapter 8

The Principle of Interpretation of Gravity Data Using Second Vertical Derivative Method

Ahmad Alhassan and Auwal Aliyu

Abstract

This chapter is aimed at demonstrating how the second vertical derivative method is been applied to gravity data for subsurface delineation. Satellite gravity data of some part of Northern Nigeria that lie between latitude 11°–13°E and longitude 8°–14°N obtained from Bureau Gravimetrique International (BGI) were used for demonstration. The Bouguer graph was plotted using surfer software. The second vertical derivative graph was also plotted. Very low gravity anomalies are observed on the Bouguer map, which recommends the presence of sedimentary rocks which have low density. The result of the second vertical derivative method has improved weaker local anomalies, defined the edges of geologically anomalous density distributions, and identified geologic units. This is a clear implication that the second vertical derivative is very important in subsurface delineation.

Keywords: gravity data, second vertical derivative, interpretation

1. Introduction

Gravity method is one of the geophysical methods widely used in environmental, engineering, archeological, and other subsurface investigations. The gravity method measures the difference in earth's gravitational field at different locations by tools known as gravimeters. Recent developments in observation, processing, and data analysis due to technological improvement added the efficiency and sensitivity of the gravity method, which made it more applicable to a wider range of problems. Airborne gravimeters have the potential to recover a precise gravity field at any place. Recent advancements in detailed aircraft positioning with global positioning system (GPS) carrier phase data have stretched the use of airborne measurement practice to land as well as overwater surveys. To date, large areas of the earth remain unmapped because of the limitations of land and marine surveys. BGI contributes to the recognition of derived gravity products with the aim of providing significant information about the Earth's gravity field at worldwide or provincial scales. Their products used mostly by scientists are the World Gravity Maps and Grids (WGM), which signify the first gravity anomalies computed in spherical geometry considering a realistic Earth model. With long-range aircraft, nearly all the Earth is accessible to airborne surveying. Improvements in hardware, software, and survey methodology continue to lower the overall error budget for airborne gravity. Even though there is no method universally agreed for evaluating data accuracy and anomaly resolution being a function of the speed of aircraft, reported RMS error and resolution indicates the technique's accuracy [1]. The gravitational field strength is directly proportional to the density of subsurface materials. This gives gravity anomalies that correlate with source body density variations. Positive gravity anomalies are connected to shallow high-density bodies and negative anomalies are connected to low-density bodies. Potential field anomaly maps present the effects of shallow (residual) and deeper (regional) geological sources. Therefore, the main issue in potential field data interpretation is the separation of anomalies into two components [2]. Consequently, in order to produce meaningful results, potential field datasets generally need many processing techniques that are in accordance with the nature of the geology of the study area. Numerous commercial software packages (e.g. Geosoft Oasis Montaj, GeosystemWinGlink, MagPick, and IGMAS) are commonly used for the analysis of potential field datasets by employing some of the methods stated above.

Quite a lot of graphical and empirical methods have been established for the interpretation of gravity anomalies that are caused by simple bodies [3]. The derivatives have a tendency to expand near-surface structures by increasing the power of the linear dimension in the denominator. This is because the gravity effect differs inversely as the square of the distance, and the first and second derivatives vary as the inverse of the third and the fourth power, respectively, for three-dimensional structures.

The second vertical derivative is frequently employed in gravity interpretation for isolating anomalies and for upward and downward continuation. This chapter aims to emphasize the advantage of using the second vertical derivative on gravity data for subsurface delineation. The second derivative is very important for gravity interpretation due to the fact that the double differentiation with respect to depth tends to emphasize the smaller, shallower geologic anomalies at the expense of larger, regional features [4]. Micro-gravimetric and gravity gradient surveying methods can be applied for the detection and delineation of shallow subsurface cavities and tunnels [5]. The second vertical derivative method is very important in edge location and edges are thought to contain most of the two mineralized ore deposits [6].

2. Theory

The gravity method is governed by Newton's law of universal gravitation and Newton's law of motion [7]. Newton's law of universal gravitation states that "the force of attraction between two bodies of known mass is directly proportional to the product of their masses and inversely proportional to the square of the separation between their centers of mass." This is expressed as.

$$F = \frac{G \times M \times m}{R^2} \qquad (1)$$

where F = the force of attraction between the masses, G = constant known as universal gravitation, M and m = respectively the masses of particles 1 and 2, and R = distance between the two masses.

Newton's second law of motion states that "the rate of change of momentum of a body is directly proportional to the applied force and takes place in the direction of the force."

Newton's second law can be expressed mathematically as follows:

$$F = Ma = Mg \qquad (2)$$

where F is the force of attraction between bodies, M is the mass of the body, and g is the acceleration due to gravity.

In differential form, Newton's second law can be stated as:

$$F = \frac{dp}{dt},$$ (3)

Therefore,

$$F = \frac{d(mv)}{dt}$$ (4)

where F = the applied force, dp = change in momentum, and dt = change in time [7].

The necessary characteristics of the gravity method can be explained in terms of mass and acceleration as illustrated in Newton's second law. The mass distribution and shape of an object are linked by the object's center of mass [8].

Equating (1) and (2), it implies that

$$g = \frac{G \times M}{R^2}$$ (5)

The gravitational potential at a point in a particular field is the work done by the attractive force of M on m as it moves from zero to infinity. The concept of the potential helps in simplifying and analyzing certain kinds of force fields such as gravity, magnetic and electric fields.

Eq. (7) represents the force per unit mass, or acceleration, at a distance r from P, and the work necessary to move the unit mass a distance (ds) having a component dr in the direction is given in Eq. (6).

$$v = Gm \int_{\infty}^{R} \frac{dr}{r^2}$$ (6)

$$v = \frac{Gm}{r}$$ (7)

where v = the work used in moving a unit mass from infinity to the point in question, m = unit mass at point P, and r = distance covered by the masses.

The gravity anomaly is the difference in values of the actual earth gravity (gravity observed in the field) with the value of the theoretical homogeneous gravity model in a particular reference datum [8].

$$\partial g_B = 2\pi Gph$$ (8)

$$g_B = g_F - \partial g_B$$ (9)

Previously, authors determined the density value at research locations based on the statement that the Bouguer anomaly can be expressed as an equation of the form of "y = mx + b" as.

$$g_{obs} - g_N + 0.3086 \, h = (0.04193 \, h - TC)p + BA$$ (10)

The units for g are cm/s^2 in the c.g.s system and are commonly known as Gals, where the average acceleration of gravity at the earth's surface is 980 Gals. Most realistic gravity studies involved variations in the acceleration of gravity ranging from 10^{-1} to 10^{-3} Gals, so most workers use the term milliGal (mGal). In some detailed work involving engineering and environmental applications, workers are dealing with microGal (μGal) variations.

3. Second vertical derivative

Many researchers [4, 8, 9] have given a thorough picture of the second derivative method of interpretation of gravity and have shown how this method is in effect for the detection of small irregularities in gravity anomalies and thus useful for supposing minute underground mass distribution that cannot be overlooked by the ordinary method. It is fascinating to note that the method is justified only on data of high accuracy [10].

The SVD of the vertical component of gravity, g_z, can be calculated in the spatial domain from the horizontal gradients by using Laplace's equation [9].

$$\frac{\partial^2 g_z}{\partial z^2} = -\left(\frac{\partial^2 g_z}{\partial x^2} + \frac{\partial^2 g_z}{\partial y^2} \right) \tag{11}$$

For an anomaly extended along the y-axis, the SVD can be approximated by the second horizontal derivative of the gravity data along the x-axis in Eq. (13).

$$\frac{\partial^2 g_z}{\partial z^2} \approx -\frac{\partial^2 g_z}{\partial x^2} \tag{12}$$

In the wave number domain or Fourier domain, the SVD is usually calculated by using the following Eq. (14) [11].

$$\frac{\partial^2 g_z}{\partial z^2} = F^{-1}\left(|k|^2 G_z \right) \text{ with } |k|^2 = k_x^2 + k_y^2 \tag{13}$$

The second vertical derivatives are the measure of curvature where large curvature is connected to shallow anomalies. It is frequently used to enhance localized subsurface features, that is, weak anomalies due to the sources that are shallow and limited in-depth and lateral extent [10].

4. Materials and methods

Qualitative interpretation of the gravity data was performed by applying second vertical derivative methods as filters to find edges of the source of gravity anomaly around the study area.

The materials used for the research include the following:

- Work station (computer)

- Surfer software

- Satellite gravity data

5. Source of data

The data used were obtained from Bureau Gravimetrique International (BGI). The survey was carried out in conjunction with IAG international Gravity field service. The main job of BGI is to gather all gravity measurements (relative or absolute) and pertinent information about the gravity field of the Earth on a worldwide basis, compile and authenticate them, and store them in a computerized database in order to redistribute them on request to multiple users for scientific applications. BGI produces the most precise information available on the Earth's gravity field at short wavelengths today and is very complementary to airborne and satellite gravity measurements. The satellite gravity data were recorded in digital layout (X, Y, and Z). X, Y, and Z represent the longitude, latitude, and Bouguer anomaly of the study area, respectively. Corrections such as drift, earth-tide, elevation and terrain, latitude and were applied on the gravity data by the Bureau Gravimetrique International (BGI).

6. The study area

The area is underlain by rock and younger sediments of the Chad formation. The Chad Basin lies within a vast area of Central and West Africa at an elevation between 200 and 500 m above sea level and covers 230,000 km². The basin is referred to as an interior sag basin, due to a sagging episode that has affected it before the onset of continental separation during which a rift system junction was formed providing an appropriate site for sedimentation. Therefore, it lies at the junction of basins, (comprising the West African rift), which becomes active in the early Cretaceous when Gondwana started to split up into component plates. Sediments are mainly continental, sparsely fossiliferous, poorly sorted, and medium-to-coarse grained, feldspathic sandstones called the Bima Sandstone. Both geophysical

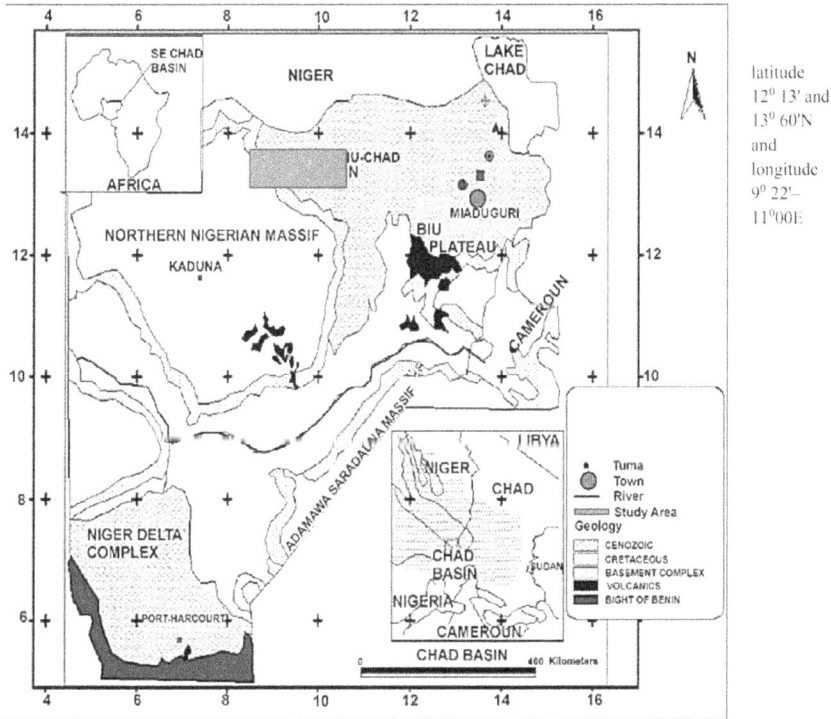

Figure 1.
Nigerian map showing Chad basin area.

and geological interpretations of data suggest a complex series of Cretaceous grabens extending from the Benue Trough to the southwest (**Figure 1**) [12].

7. Methodology

The target of the gravity method is the determination of important facts about the earth's subsurface. One can just study the grid of gravity values for the determination of the lateral location of any gravity variations or perform a more thorough analysis to calculate the nature (depth, geometry, density) of the subsurface structure that caused the gravity variations. To determine the latter, it is usually necessary to distinguish the anomaly of interest (residual) from the remaining background anomaly (regional).

Gravity data were analyzed and interpreted using the second vertical derivative method. Performance of horizontal and vertical derivatives was evaluated using synthetic data. SVD calculation using 2D was applied to identify fault structure. The first derivatives of the horizontal component (dg/dx and dg/dy) were calculated in the excel software. Then, the second derivatives ($\frac{\partial^2 g}{\partial x^2}$ and $\frac{\partial^2 g}{\partial y^2}$) were calculated in the same software.

These are used to find the second vertical derivative ($\frac{\partial^2 g}{\partial z^2}$) using relations in Laplace's equation as mentioned in Eq. (12). The second vertical derivative values are then gridded in surfer software. The grids are then used to plot the contour map known as the SVD map in **Figure 3**.

The second vertical derivative map is expected to remove the effect of regional trends. The edge of the residual anomaly was then observed, which is seen on zero contours. This will help to predict the anomaly in the map with its position.

8. Results and discussion

The application of the second vertical derivative method in this research has yielded results. **Figures 2** and **3** show the results of the Bouguer anomaly distribution, application of second vertical derivative to the Bouguer values. The data obtained from Bureau Gravimetric International (BGI) were used throughout the work. The data had undergone Bouguer correction already. While the Bouguer result shows the density variation across the study area, the second vertical derivative graph shows the major fault zone. These results are presented as follows:

8.1 Bouguer graph

The data were converted from excel to data file format using grapher software. Its grid and contour maps were obtained using surfer software. The Bouguer map in **Figure 2** indicates that the Bouguer anomaly of the study area varies from −60 to −4 mGal. Low gravity anomalies are observed in the map with its minimum value appearing in around 11.69°N–11.85°N, 11.2°E–11.33°E, and 11.0°N–11.2°N, 11.5°E–11.7°E. Meanwhile, Bouguer gravity anomalies are maximum around 11.70° N–11.80°N, 12.73°E–12.84°E.

8.2 Second vertical derivative

The Bouguer data were then filtered using a second vertical derivative on surfer software to produce a new grid and subsequently, the contour map was plotted as shown in **Figure 3**. The second vertical derivative map shows the distribution of the second-order vertical derivative in mGal/m^2. The value ranges from $-320,000,000$ mGal/m^2 to $100,000,000$ mGal/m^2. The zero values show the edges of the deeper feature. As defined earlier, the portion with less density is considered to contain more sediment. The map shows that density distribution is decreasing inward between 8°N to 10.5°N and 11°E to 13°E. The same thing happens between 12.5°N to 13°N and 11.2°E to 12°E. The zero contours can be observed throughout the map, which means that there are so many boundaries or edges in the area.

The second vertical derivative map in **Figure 3** shows that the "polarity" of the anomaly can still be recognized, that is, the low density in the Central part relative to its surroundings. The second vertical derivative method of gravity anomaly illustrates the amplitude of gravity anomaly that is triggered by fault structure that gives the impression of residual anomaly. The map of the second vertical derivative method in this study shows that the method is useful in enhancing weaker local anomalies, edges of geologically anomalous density distributions were defined, and geologic units are identified. The second vertical derivative is interested in near-surface anomalous effect at the expense of the effects that are of deep origin. This study embraced the use of the second vertical derivative because of its tendency to emphasize local anomalies and isolate them from the local background, which can be seen in **Figure 3**. When compared to the Bouguer map in **Figure 2**, it can be observed that the calculation of gradients has boosted refined features of gravity data that else cannot be noticed visually from the original data. High gradients observed in the middle of the map can be connected to the high contrast of the subsurface physical properties and vice versa [11].

Gradients, and also their magnitude, are commonly engaged to delineate boundaries of anomalous sources. The map produced extended zero contours,

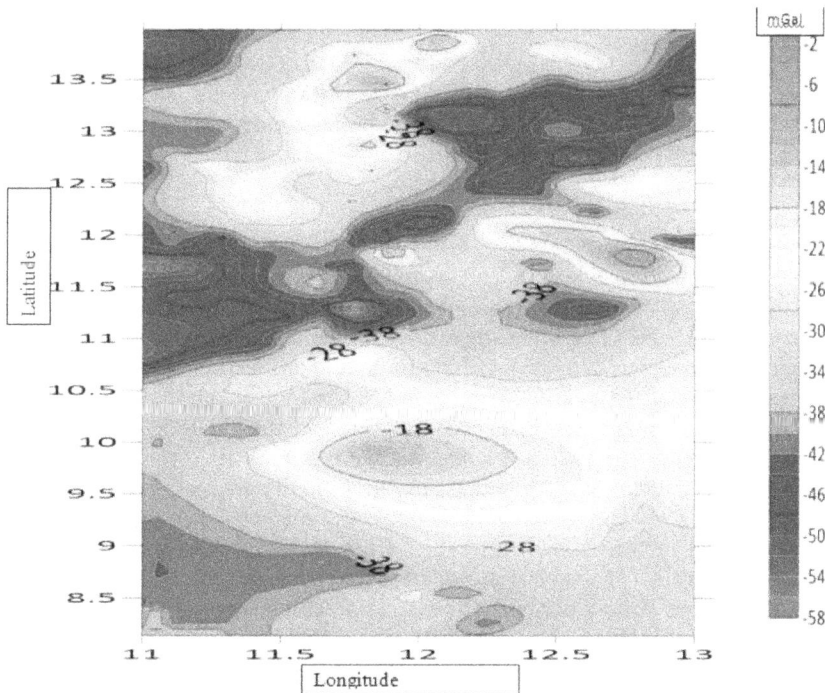

Figure 2.
Bouguer gravity map.

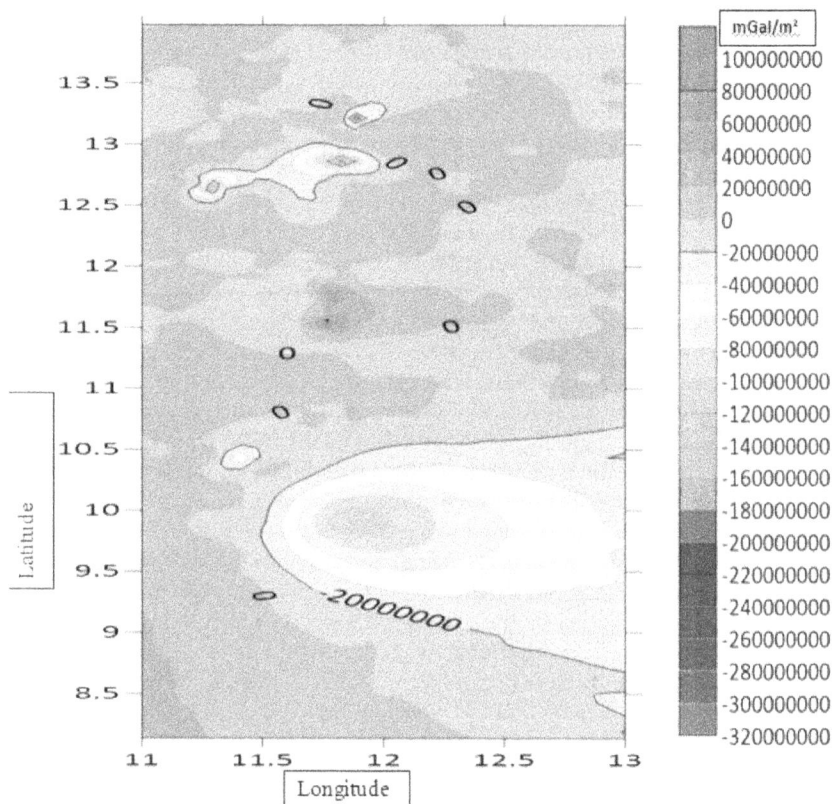

Figure 3.
svd graph showing zero edges.

which corresponds to the edges of local geologically anomalous density distribution structures. The quantity zero mGal/m² coincides with most of the lithological boundaries, when compared with the major geologic contacts.

9. Conclusion

In this chapter, the use of the second vertical derivative is described as one of the efficient methods used for enhancing weaker local anomalies, defining the edges of geologically anomalous density distributions, and identification of geologic units. Satellite gravity data of a particular place in Nigeria were acquired from Bureau Gravimetrique International (BGI). The data that had already undergone Bouguer correction were used to plot the Bouguer map of the study area, which shows that the place is a semidentary basin because negative gravity anomalies are observed throughout the area. The second vertical derivative map was then plotted to emphasize local anomalies and isolate them from the local background, which can be seen in **Figure 3**. The map has shown areas that have lower and higher anomalies of deeper sources. Boundaries of the anomalies are also observed. On the second vertical derivative maps, the "polarity" of the anomaly can still be identified, that is, the low density in the Central part relative to its surroundings. The second vertical derivative method in this study shows that it is useful in enhancing weaker local anomalies, defining the edges of geologically anomalous density distributions, and identification of geologic units. Boundaries are better delineated by the second vertical derivative method with oscillations between the minimum and maximum (extremum) values through each density contrast transition (**Figure 3**). It is important to note that the second vertical derivative method is justified only on data that has high accuracy [10].

Author details

Ahmad Alhassan[1*] and Auwal Aliyu[2]

1 Faculty of Sciences, Department of Physics, Federal University of Kashere, Gombe State, Nigeria

2 Faculty of Sciences, Department of Physics, Federal University Gashua, Yobe State, Nigeria

*Address all correspondence to: ibnalhassan2010@gmail.com

IntechOpen

References

[1] Geodesy IA. Journal of Geodesy, The Geodesist Hand Book. 2016

[2] Yunus Levent Ekinci and Yigitbas Erdinc. Interpretation of gravity anomalies to delineate some structural features of Biga and Gelibolu peninsulas and their surroundings (North-West Turkey). Geodinamica Acta. 2015;27(4): 300-319

[3] Essa KS. A fast interpretationmethod for inversemodeling of residual gravity anomalies caused by simple geometry. Journal of Geological Research. 2012. Article ID 327037, 10 pages

[4] Elkins TA. The second derivative method of gravity interpretation. Geophysics. 1951;16(1):29-50

[5] Wahyudi EJ, Kynantoro Y, Alawiyah S. Second vertical derivative using 3-D gravity data for fault structure interpretation. In: International Conference on Energy Sciences. IOP Publishing IOP Conf. Series: Journal of Physics: Conf. Series; 2016. p. 877

[6] George H, Menyeh A, Wemegah DD. Qualitative interpretation of aerogravity and aeromagnetic survey data over the sourth Western part of the Volta River basin of Ghana. International Journal of Scientific and Technology Research. 2015;4(04):23-30

[7] Ofoha CC, Emujakporue G, Ekine AS. Structural evaluation of bouguer gravity data covering parts of Southern Niger Delta, Nigeria. Archives of Current Research International. 2018; 13(3):1-18

[8] Justia M, Hiola MFH, Nur Baiti Febryana S. Gravity anomaly to identify Walanae fault using second vertical derivative. Journal of Physics: Theories and Applications. 2018;2:36-42

[9] Aku MO. Application of second vertical derivative analytical method to bouguer data for the purpose of delineation of lithological boundaries. Science World Journal. 2014;9(3):27-32

[10] Sumintadireja P, Dahrin D, Grandis H. A note on the use of the second vertical derivative (SVD). Journal of Engineering Science and Technology. 2018;50(1):127-139

[11] Putra MI, Hidayah AN, Septiana LAES, Supriyana E. Gravity analysis to estimate the mud stream direction in porong by applying the delaunay approach. In: IOP Conference Series: Earth and Environmental Science. 2019. p. 311

[12] Chukwunonso OC, Godwin OA, Kenechukwu AE, Ifeanyi CA, Emmanuel IB, Ojonugwa UA. Aeromagnetic interpretation over Maiduguri and environs of Southern Chad Basin, Nigeria. Journal of Earth Sciences and Geotechnical Engineering. 2012;2(3):77-93

www.ingramcontent.com/pod-product-compliance
Lightning Source LLC
Chambersburg PA
CBHW081224190326
41458CB00016B/5676